STORAGE BATTERIES

STORAGE BATTERIES

Including Operation, Charging, Maintenance and Repair

G. SMITH C.ENG., M.I.E.E.

Technical Information Officer
Electric Power Storage Ltd.

Pitman Publishing

First published 1964
Reprinted with corrections 1968
Second edition 1971

SIR ISAAC PITMAN AND SONS LTD.
Pitman House, Parker Street, Kingsway, London, WC2B 5PB
P.O. Box 6038, Portal Street, Nairobi, Kenya

SIR ISAAC PITMAN (AUST.) PTY. LTD.
Pitman House, Bouverie Street, Carlton, Victoria 3053, Australia

PITMAN PUBLISHING COMPANY S.A. LTD.
P.O. Box 11231, Johannesburg, S. Africa

PITMAN PUBLISHING CORPORATION
6 East 43rd Street, New York, N.Y. 10017, U.S A.

SIR ISAAC PITMAN (CANADA) LTD.
Pitman House, 381–383 Church Street, Toronto, 3, Canada

THE COPP CLARK PUBLISHING COMPANY
517 Wellington Street, Toronto, 2B, Canada

ISBN: 0 273 36087 6

MADE IN GREAT BRITAIN AT THE PITMAN PRESS, BATH
(T.1252)

PREFACE TO THE SECOND EDITION

DESPITE all the wonders and scientific achievements of our age, electricity from the power stations cannot be stored for use at a later date. It is instant power in the sense that it must be used immediately it is produced.

There are many demands for portable electrical power, remote from the mains supply. This type of equipment is growing steadily and in almost all cases is powered by a wide variety of batteries, providing instant electricity at any time.

Since the publication of the first edition of this book in 1964, many interesting developments have taken place in the field of battery operated epuipment and appliances.

A market has been created for a new range of small rechargeable batteries to provide power for a wide variety of electrically operated toys and other equipment known as cordless appliances. These include shavers, electric drills, hedge trimmers, radios, television sets, tape recorders, toothbrushes, walkie-talkie sets, lawn mowers, etc. The lead-acid and nickel-cadmium batteries used for these appliances provide more power and longer life than the Leclanché (primary) dry battery which has been so successful for many years on appliances with low power demands.

Lead-acid batteries and nickel-cadmium batteries have been specially designed for these applications to provide reliability and long life without the necessity for attention or maintenance other than plugging in to a mains supply for recharging the battery.

The continued expansion of the world car population has increased the hazards of air pollution from engine exhaust gases in all large cities, up to the point where alternatives to the internal combustion engine have been explored.

The obvious first choice was the electric car powered by lead-acid or nickel-cadmium batteries. Many prototype passenger cars, buses and small commercial vehicles have been produced in the United States of America, Japan, Germany, Great Britain and elsewhere, using these types of batteries.

Whilst it is true that an electric passenger car powered by these batteries would satisfy 60 to 70 per cent of car owners who travel about 50 kilometres (30 miles) each working day, the trials have demonstrated that their range is limited by the battery whilst the time of 6 to 8 hours to recharge the batteries is far too long to compete with the performance of an i/c engine car with its long range, faster speed and quick refuelling system.

Other developments in lead-acid car batteries include the introduction of water-activated batteries which merely require the addition of water in place of sulphuric for putting into service.

Some of the major battery companies are also introducing battery cases and lids in polypropylene or similar plastic materials, in a variety of bright colours. Apart from improving the old "black box" image, these new containers are lighter in weight and tougher than the conventional hard rubber cases.

All the above new applications and batteries are described in Chapter 9.

In the past ten years many new systems have been under development with optimistic claims for higher energy outputs and performance compared to conventional storage batteries. An enormous amount of money and research has been put into the development of fuel batteries. It is now clear that the bright commercial future forecast for this system was over-optimistic, although short term it has been highly successful in space vehicles and similar projects where manufacturing cost is not important.

The principle of operation, performances and future of fuel batteries, metal-air and other interesting systems are described in Chapter 11.

All units shown in the book have been expressed in metric with the equivalent Imperial Unit shown alongside.

G. S.

PREFACE TO THE FIRST EDITION

THE development of modern industry and the desirable increase in national productivity depend to a large extent on an expanding and efficient electricity supply industry.

To this end, new and larger power stations, of conventional and nuclear design, are being built to satisfy increased electrical load demands, which have doubled in the last ten years.

The electricity supply and storage battery industries have been closely associated since the earliest days of direct-current power stations. Then, very large storage batteries were an essential part of the power plant, providing electrical power which was switched on to the system either to supplement or take over generator loads during periods of heavy and light working.

Changes from d.c. to a.c. generation and the growth of the national grid system have altered the demands made on storage batteries. Batteries of very large capacity are no longer required, but because of the complexities of high-voltage a.c. generation and distribution many more smaller batteries are used for a wide variety of duties.

Improved standards of living have created a large increase in the demand for telephones and all kinds of motor-cars and road vehicles —none of these can work without a storage battery. Other forms of transport, by air, rail and sea, rely on batteries for vital standby power. The replacement of steam locomotives by Diesel and Diesel-electric locomotives has increased the demands on the battery, which must now supply very high electrical power for starting large Diesel engines, in addition to lighting loads.

Extensive building projects for new towns, schools, hospitals, large stores, places of entertainment, etc., demand storage batteries for maintaining emergency services during power failures. Industrial efficiency is closely related to improved standards of material-handling relying largely on battery-electric powered trucks for transport and stacking duties.

The adaptability and reliability of storage batteries as sources of electrical power have made them an essential part of most industries throughout the world. Yet, in spite of their widespread use, batteries remain to many people the black box of mystery and magic. The black box conceals not one but many designs, each with some special built-in characteristic to enable a particular battery to give optimum performance for a particular duty.

For each operation or duty there is the right type of battery, and correct charging and maintenance procedure to ensure the longest possible trouble-free life. A little knowledge of how a storage battery works, and the simple maintenance required for maximum service, will almost certainly save time, money and frustration.

This book has been written in the hope that it will provide guidance in the choice, operation and maintenance of the various types of battery available.

G. S.

ACKNOWLEDGEMENTS

I WISH to thank my colleagues, Mr. W. A. Lees, Mr. W. Lord and Mr. A. Urmston, for their assistance in reading parts of the manuscript.

I am indebted to Mr. V. A. Lord and Mr. J. Prest for their comments on the complete manuscript, and to Mr. V. A. Lord for his advice on the section dealing with stationary batteries.

Acknowledgement is made to the following firms and organizations for their kind permission to use various information and illustrations: Britannia Batteries Ltd., Central Electricity Generating Board, City and Guilds of London Institute, Conveyancer Fork Trucks Ltd., Electric Power Storage Ltd. (Dagenite Exide and divisions), Enfield Automotive, Greenwood & Batley Ltd., Lansing Bagnall Ltd., Legg (Industries) Ltd., Nife Batteries Ltd., Smith's Delivery Vehicles Ltd., Westinghouse Brake & Signal Co. Ltd.

G. S.

CONTENTS

PRINCIPAL SYMBOLS AND ABBREVIATIONS

Quantity	Symbol	Unit	Abbr.
Capacity; (rated or nominal)	C	ampere-hour	Ah
Current	I	ampere	A
		milliampere	mA
Electromotive force (e.m.f.) .	E	volt	V
Energy	W	watt-hour	Wh
		kilowatt-hour	kWh
Frequency	f	cycle per second	c/s
Power	P	watt	W
		kilowatt	kW
Resistance	R	ohm	Ω
		megohm	$M\Omega$
Time	T	second	sec
		minute	min
		hour	hr
Voltage	V	volt	V

Specific Gravity. Differences between readings are often expressed in *points.* Thus $1 \cdot 280 - 1 \cdot 250$, or $0 \cdot 030$, is 30 points.

INTRODUCTION

Primary and Secondary Cells

VOLTAIC CELLS store chemical energy which is converted to direct-current electrical energy during discharge. They are divided into two classes, the *primary cell* and the secondary cell, accumulator or *storage cell*.

The chemical reactions in storage cells are reversible: after discharge such a cell can be restored to its original chemical condition by passing an electric current through it in the direction opposite to that of discharge.

For all practical purposes the chemical reactions in a primary cell are irreversible: the generation of electric current consumes materials which cannot be replenished by recharging.

A *battery* consists of two or more voltaic cells connected together. The storage batteries described in this book are the types most widely used for commercial purposes, namely *lead-acid* batteries and alkaline batteries of the *nickel-cadmium* and *nickel-iron* types.

CHAPTER 1

FUNDAMENTALS OF THE STORAGE BATTERY

THE storage of electrical power remains one of the greatest challenges to modern science. For most practical purposes electricity can be used only as it is generated. When the machine generating electricity at the power station stops, the wires linking power station and consumer become "dead" and no longer pass electricity.

There are, however, various systems and devices for converting electrical energy to some convenient form of stored energy for reconversion to electrical energy when required. Recently, hydro-electric pump storage stations have been built which utilize spare electrical power for pumping water to high-level reservoirs. The stored water is used later to drive turbines for producing electricity during periods of peak loading.

Since the earliest days of electrical power supply, storage batteries have been widely used to meet lighting and power demands during failure of mains supply, or to provide electrical power for equipment used away from the main electricity supply.

In the storage battery, electrical energy is stored as chemical energy which is converted to electrical energy, instantly and silently, and without the necessity of any moving parts, merely by closing a switch. The lead-acid battery is the type of storage battery which is most widely used, and the reasons for this will be explained in later chapters.

The necessity for storing electrical power may appear perplexing to the layman, who may have heard on many occasions that electricity exists everywhere and in everything. This is true, but electricity exists in atoms containing positive and negative electrical charges, which under normal conditions are balanced so that there is no movement of electrons within the material to produce an electric current. An electric current is therefore produced when the balance of electrical charges within atoms of the material is upset by some outside influence causing electrons to pass from one atom to the next.

With certain substances such as metals the electrons move freely, once their movement is started by the outside influence. This ready transfer of electrons is in fact an electric current in the metal conductor, and as many electrons are expelled at the one end as are

1

accepted at the other end of the conductor. One simple method of starting the transfer of electrons, or flow of electricity, in a conductor is to connect a source of electromotive force, such as a storage cell or battery, across the ends of the conductor.

Electrical Power from Lead-Acid Storage Batteries

The rapid development of electric lighting about 1880 and the improvements made in dynamo designs for generating direct current led to the creation of central supply stations in most cities and large towns. The new supply stations required storage batteries for—

(*a*) Maintaining the small outputs at night time when the engines which drove the generators were shut down.
(*b*) Maintenance of supply during breakdowns.
(*c*) Load sharing during periods of maximum demand.

By 1900, the widespread demand for batteries for such purposes as lighting, propulsion of road vehicles, boats and tramcars, operation of telephone exchanges, etc., prompted the growth of various battery works and the foundation of the battery industry.

In 1911 the invention in America of the electric starting system for motor-cars created a demand for many more car batteries of increased power. Today the demand for motor-car and vehicle batteries amounts annually to several millions, and accounts for more than 80 per cent of the lead used in the battery industry.

From those early days the demand for batteries for all kinds of duties has steadily increased, whilst the last twenty years have seen a rapidly growing demand for batteries for specialized uses.

The electrical devices and equipment used in the Second World War required portable battery power in unprecedented designs and numbers. Light-weight batteries were required for aircraft, portable radios, telephones and transmitters. High-capacity heavy-duty batteries were required for tanks, army vehicles and assault craft. Special batteries of unique design were made for the destruction of enemy magnetic mines, for submarines of conventional and miniature size. Safety on land and sea was served by batteries designed to provide lighting in air-raid shelters, hospitals, public buildings, ships and lighthouses. Lifeboats carried radio transmitters worked by batteries designed specifically for that duty. In the munition factories, down the mines, on the docks, thousands of battery-propelled trucks of all types worked round the clock, speeding up the manufacture and transport of vital munitions and supplies.

Since the war the change-over and expansion of industry, together with an ever increasing rise in the standard of living, have created a tremendous demand for electrical power. Attempts to satisfy this have been made by the building of numerous power stations, including several using nuclear fuel. Even these huge power plants require storage batteries for standby duties during an emergency or breakdown. The maintenance of essential services by battery power during an emergency is particularly vital in the nuclear power stations.

Stationary batteries have also played a very important part in the development of automatic telephone exchanges, and the British Post Office operates more battery power than any other user. Many more batteries will be required for the vast conversion and expansion schemes which are still proceeding in this field.

The need for portable or localized power has also greatly increased. The number of battery-operated industrial trucks and vehicles in Great Britain has increased from 25,000 to 90,000 in the last 10 years: the number of motor vehicles, each requiring a battery, multiplies year by year.

Improvements in Lead-Acid Storage Batteries

In all applications the battery has proved to be reliable and has frequently given good performance under adverse conditions and with indifferent maintenance. Much of the success of modern batteries lies in the research carried out since the war, resulting in improvements in battery components and production of designs suited to various operating conditions.

Battery life has been extended by the use of improved grid alloys, container and separator materials. Performance under widely different temperature conditions has been improved by new process methods and blending of the oxides with new additives. The energy outputs per unit weight and volume have increased by 30 per cent in the batteries designed for traction purposes. The modern car battery gives 20 per cent more energy per pound than that of 10 years ago. Modern stationary batteries for emergency duties in power stations, large stores, etc., give $2\frac{1}{2}$ times the output for the same volume compared with those made 20 years ago.

Modern dry-charged batteries can be stored dry for a period of 2 years, and still give at least 75 per cent of nominal capacity when activated by adding sulphuric acid.

History and Development of the Storage Battery

Much of the experimental work which led to the development of the lead-acid battery was carried out by Gaston Planté in 1859. Earlier, other scientists, the most notable being Gautherot, De la

Rue, Ritter, Grove, Faraday and Sinsteden, had each made some contribution to the discovery of the ideal reversible reaction, which in its practical form is the fundamental principle of the storage battery.

PLANTÉ (FORMED) PLATES

The earliest cell made by Planté consisted of two lead sheets, separated by strips of flannel, rolled together and immersed in dilute sulphuric acid. Planté discovered that, by passing an electric current through the plates for a considerable period, he was able to convert the surface of one plate to lead dioxide (positive) and the surface of the other plate to spongy lead (negative). This combination produced an electric current when the two plates were connected to an external circuit, and by alternate discharge and charge, Planté was able to increase the storage capacity of the cell. The only source of charging available to Planté was by primary battery, and his early cells took as long as a year to charge. It was not surprising, therefore, that for the first ten years following the invention, Planté cells, or accumulators as they were often called, were used only in laboratories where they were capable of supplying current in excess of that available from primary cells.

FAURE (PASTED) PLATES

In 1881, Faure, a Frenchman, and an American named Brush took out, independently, patents for pasting the surface of lead plates with coatings of lead oxides which could be readily formed, by the passage of current through the plates, into the active materials lead dioxide and spongy lead. The adherence of the active materials to the solid lead plates was poor, and the paste was easily shed from the plate surfaces.

An improved plate containing numerous holes filled with a paste of lead oxide and sulphuric acid was patented by Volckmar, also in 1881. The idea was developed by Swan and Sellon, who produced a grid for holding the paste.

In 1882 Sellon patented a modified grid using an alloy of lead and antimony, the grid mesh being designed to key the active material in the grid. This grid was to be the basic design for pasted plates used in most portable lead-acid batteries.

ALKALINE STORAGE BATTERIES

Planté and other scientists had experimented with various combinations of metal electrodes and electrolytes in their search for the perfect reversible cell. Most combinations or couples had no practical value and were merely laboratory curiosities.

After much experimental work the only other type of cell which was comparable in commercial value with the lead-acid cell was invented by Edison of America and Jungner of Sweden. In 1900 Jungner patented the nickel-cadmium-alkaline cell, and a year or so later Edison patented the nickel-iron-alkaline cell. Both types used an alkaline electrolyte of dilute potassium hydrate, and were the prototypes of present-day nickel-cadmium and nickel-iron batteries.

Definition and Chemistry of the Storage Battery

A storage battery is a chemical device reversible in its action, which stores energy at one time for use at another. The energy stored is chemical, not electrical. Electrical energy in the form of direct-current electricity is applied to the battery during the operation termed *charging*. The electric current produces chemical changes in the battery, and the chemical energy stored in the plates is reconverted to electrical energy when the cell is *discharging*.

Types of storage battery which have most practical and commercial value are lead-acid batteries and alkaline batteries, the latter including nickel-iron and nickel-cadmium assemblies.

LEAD-ACID STORAGE CELLS

The fundamental parts of a lead-acid storage cell are two dissimilar plates, or electrodes, immersed in an electrolyte in a suitable container, namely

Positive active material .	. Lead dioxide (PbO_2)
Negative active material.	. Spongy lead (Pb)
Electrolyte . . .	. Dilute solution of sulphuric acid (H_2SO_4) in water

In a fully charged healthy cell the positive active material is dark chocolate colour, and the negative active material slate-grey colour.

Lead plates *similar* in chemical composition and immersed in dilute sulphuric acid produce *no* chemical reactions or electric current when connected to an external circuit.

The most elementary form of commercial lead-acid cell, consisting of a single pair of plates in a plastic (polystyrene) box, is shown in Fig. 1.1. This type is designed to supply small discharge currents, and is used mainly for bell circuits, alarms and laboratory work.

Although the lead-acid battery suffers the handicap of utilizing one of the heaviest of metals—lead—it is still the most widely used of the storage battery devices. Some of the characteristics which have contributed to its success are worth noting—

(*a*) The lead-acid battery has the lowest initial and operating costs of the various types of storage battery. These include alkaline batteries, (nickel-cadmium and nickel-iron) and silver-oxide-zinc.

(*b*) Its voltage on discharge is the highest of all the reversible combinations used.

(*c*) It uses comparatively cheap and plentiful materials.

(*d*) The completely reversible chemical reactions produce little physical change in the plates.

(*e*) It can operate satisfactorily over a wide range of temperature from approximately $-18°$ to $43°$ C $(0°$ to $110°$ F).

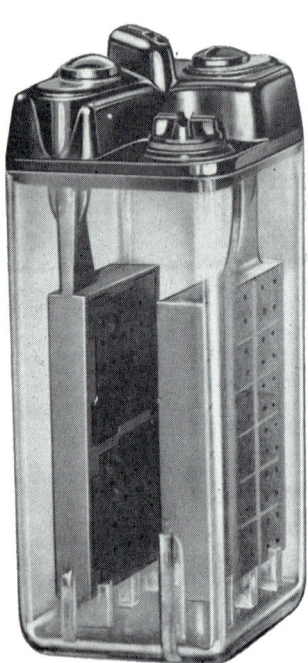

FIG. 1.1. LEAD-ACID CELL OF SIMPLE CONSTRUCTION OF ONE POSITIVE AND ONE NEGATIVE PLATE IN A POLYSTYRENE CONTAINER

CHEMICAL REACTIONS OF LEAD-ACID BATTERIES

In general, a storage battery must be given a charge before it can function, and this is carried out by connecting a suitable low-voltage d.c. supply across its terminals for a certain number of hours. There are, however, some batteries which can be activated merely by adding acid. These batteries, usually of the automotive type, are fitted with plates which have been specially processed or "dry charged" (*see* Chapter 6).

When a battery is fully charged the chemical changes taking place within the cell are complete. The positive active material has been converted to lead dioxide (PbO_2), and the negative to spongy lead (Pb), in contact with the electrolyte of dilute sulphuric acid (H_2SO_4).

DISCHARGING

When the battery is discharged by connecting a conductor across its terminals, a current will flow in the external circuit from the positive to the negative terminal. Current also flows inside the battery between plates of opposite polarity by way of the conducting sulphuric acid solution.

It is in the dilute sulphuric acid (the electrolyte) that very important chemical changes take place when a current passes between the battery plates. This is very different from the flow of current (electrons) in the conductor across the terminals of the battery, which leaves the conductor completely unchanged.

Electrolytes are substances which, in the liquid state or in solution, are largely dissociated into positive and negative ions, or charged particles. Thus in solution a molecule of sulphuric acid (H_2SO_4), which is electrically neutral, is dissociated into one sulphate ion (SO_4^{--}), carrying two electronic charges, and two hydrogen ions (H^+), each carrying a positive charge which is numerically equal to the charge of an electron ($1 \cdot 602 \times 10^{-19}$ coulomb).*

It is the migration of these ions to the electrodes (plates) immersed in the sulphuric acid which causes electricity to flow within the cell. Most of the chemical changes take place at the surface of the plates in contact with the electrolyte, for it is here that the ions produce chemical changes within the active material.

CELL ON OPEN-CIRCUIT

With no external circuit connected to the terminals of the cell, the two sets of ions within the electrolyte are in equilibrium and prevented from moving to the respective plates.

CELL ON DISCHARGE

When an external circuit is connected across the cell terminals the sulphate ions move to the negative plate and part with their negative charge. This produces an excess of negative charge on the plate, which is relieved by a flow of electrons into the conductor from the negative terminal to the positive terminal, that is, from a point of low potential to one of higher potential. (This is opposite to the conventional direction of electric current, which is that in which positive charges would move—if they could—in the external circuit.) The passage of surplus electrons from the negative plate to the conductor allows more sulphate ions from the electrolyte to combine with the lead to form lead sulphate ($PbSO_4$).

At the positive plate, the highly oxidized lead dioxide (PbO_2) is short of negative charge, so that it readily accepts the electrons arriving from the conductor. Hydrogen ions now move in to the positive plate from the electrolyte and combine with oxygen to

* 1 coulomb = 1 ampere-second; 60×60 coulombs = 1 ampere-hour.

form water (H_2O). This leaves some lead free to combine with the sulphuric acid to form lead sulphate and more water.

As the discharge proceeds and current continues to flow, more lead sulphate is formed, in both plates, by combination of the acid from the electrolyte. Water also is manufactured, which helps to dilute the electrolyte, and it is this progressive weakening of the electrolyte by formation of water which provides a convenient way of measuring the amount of discharge taking place. The cell is discharged when its voltage falls rapidly, and at this stage most of the active material has been converted to lead sulphate and the plates are almost identical in chemical composition.

CELL ON CHARGE

To reverse the chemical changes taking place in the cell during discharge, it is necessary to pass a current into the cell in the opposite direction to that of discharge.

The charging source must therefore have a voltage greater than that of the cell or battery to be charged. The charging source connected across the cell supplies an excess of negatively charged electrons to the negative plate and creates a shortage at the positive plate. The result is that hydrogen ions (positively charged) are attracted to the negative plate, where the hydrogen combines with the lead sulphate to form lead (Pb) and acid (H_2SO_4).

The shortage of charge produced at the positive plate results in sulphate ions being attracted, and combining with hydrogen of the water to form sulphuric acid. Some of the oxygen of the water combines with the lead of the positive plate to form lead oxide. At the negative plate the process of recombination of the hydrogen and sulphate continues as long as there is sulphate present. When the process of conversion of lead sulphate to lead is almost complete, hydrogen bubbles form at the negative plate and rise through the electrolyte.

Similarly sulphate ions react with water at the positive plate, forming sulphuric acid and leaving oxygen to react with lead to form lead dioxide. When most of the lead is converted, the oxygen appears as gas at the positive plate. The formation of hydrogen and oxygen gas at the plates is a sign that the cell is reaching the fully charged condition.

As the charge proceeds, acid which is released from the plates passes into the electrolyte and the specific gravity slowly increases. Measurement of the specific gravity of the electrolyte during the course of a charge does not give a true indication of the charged condition of the cell or battery. It is not until *gassing* commences that the stronger acid, liberated from the plates, is mixed with the weaker acid at the top of the cell. Specific gravity readings can therefore be of value only towards the end of the charge, when constancy of readings indicates that all the strong acid has been liberated from the plates and the cell is fully charged.

Although some gassing is necessary to bring into circulation the strong acid released during charge, excessive and prolonged gassing does no good, and can in time shorten the life of a battery by scouring the active materials at the surface of the plates.

Hydrolysis of the electrolyte results in loss of water which must be replaced by adding pure water from time to time. This is termed topping-up. 100 Ah of gassing will drive off about 34 cm³ of water, and the demands of a battery as regards the amount of topping-up water required, when working a known duty, are a good guide to correct charging. Excessive water usually means overcharging, whilst too little means undercharging.

The chemical reaction can be expressed as follows—

$$\underset{\text{Discharged}}{} \qquad \underset{\text{Charged}}{}$$

$$PbSO_4 + PbSO_4 + 2H_2O \rightleftarrows PbO_2 + Pb + 2H_2SO_4$$

Lead sulphate Positive	Lead sulphate Negative	Diluted electrolyte	Lead dioxide Positive	Lead Negative	Sulphuric acid

The arrows are used instead of an equals sign ($=$) to indicate that the reaction is reversible.

It will be appreciated that the normal working of a battery produces lead sulphate at both positive and negative plates, which are reconverted to their original condition by charging. *Sulphated* is the term usually applied to a battery which has been abused by undercharging, or leaving in a discharged condition for long periods, when the plates become excessively sulphated, lose porosity and develop a high resistance. In this abnormal condition the sulphated plates will not usually accept a charge, but some of the less severe cases can be recovered by special treatment.

Fig. 1.2 is a diagrammatic illustration of the various stages in the complete cycle of discharge and charge of a lead-acid storage battery.

Classification of Storage Batteries

There are many types of battery designed to give optimum performance under different operating conditions. Battery performance can be a measure of one or more characteristics whose relative importance will vary according to the nature of the duty. For some duties, life or portability, for some, maximum output per unit

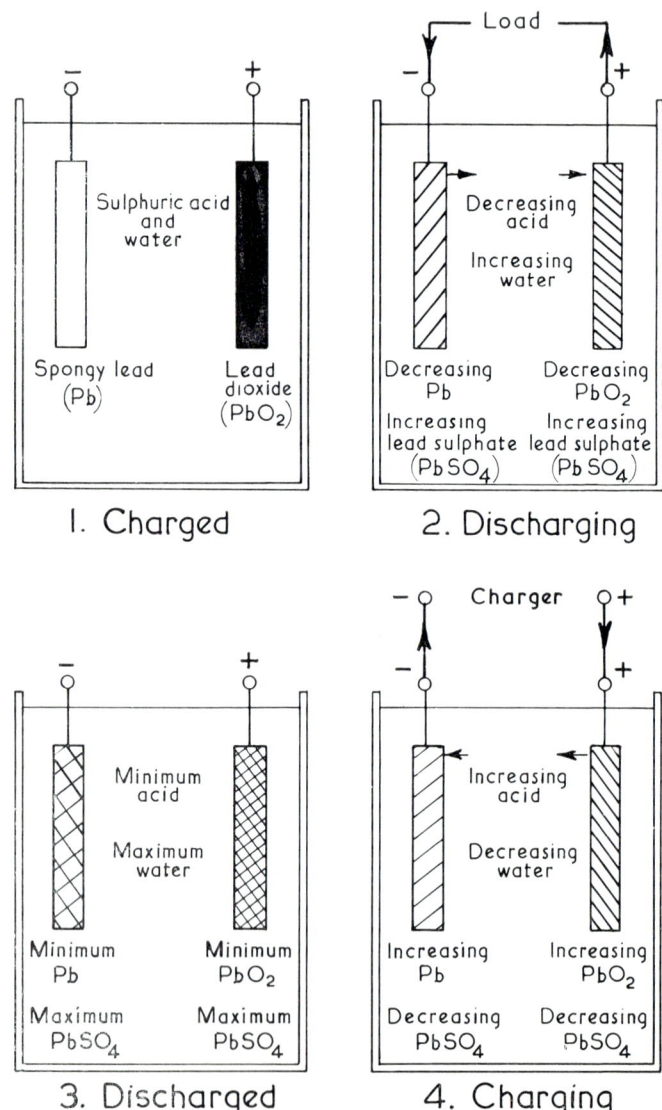

Sulphuric acid
and
water

Spongy lead
(Pb)

Lead
dioxide
(PbO₂)

1. Charged

Decreasing
acid

Increasing
water

Decreasing
Pb

Decreasing
PbO₂

Increasing
lead sulphate
(PbSO₄)

Increasing
lead sulphate
(PbSO₄)

2. Discharging

Minimum
acid

Maximum
water

Minimum
Pb

Minimum
PbO₂

Maximum
PbSO₄

Maximum
PbSO₄

3. Discharged

Increasing
acid

Decreasing
water

Increasing
Pb

Increasing
PbO₂

Decreasing
PbSO₄

Decreasing
PbSO₄

4. Charging

Fig. 1.2. Chemical Reactions within the Lead-Acid Cell during Discharge and Charge

weight or volume, and for others, low operating costs may be all-important characteristics.

Batteries may be divided roughly into two classes in relation to the type of duty they have to satisfy. Batteries which are to be operated at various locations, or carried about in the course of operation either by hand or in the vehicle of which they form a component part, are called *portable batteries*. Batteries which are to be static, or located in a fixed place during their life, are called *stationary batteries*.

Portable batteries include

Automotive batteries (car and motor vehicles)
Traction batteries (vehicles, industrial trucks, etc.)
Submarine, and marine batteries
Miners' hand- and cap-lamp batteries
Diesel locomotive starting batteries
Train batteries
Motorcycle and scooter batteries
Radio and small laboratory batteries
Aircraft batteries

Stationary batteries include

Power station batteries
Telephone exchange batteries
Emergency lighting batteries
Engine starting batteries
Fire and alarm batteries
Clock batteries
Laboratory batteries

CHAPTER 2

STORAGE BATTERY CONSTRUCTION AND DESIGN

IN the previous chapter it was explained that for operational purposes the storage battery could be conveniently classified as portable or stationary.

Portable Lead-Acid Battery Construction

The Faure or pasted plate is used in all lead-acid batteries for portable applications on account of its higher capacity per unit weight and volume compared to that of the Planté plate. The main difference between the two types is that the active material of the pasted plate is in the form of a paste held securely in an antimonial-lead grid or frame, whilst the active material of the Planté positive plate is derived from the lead of the plate itself by electrochemical action during charge.

Pasted plates vary considerably in size and thickness, the thinnest and smallest being used where portability combined with a high ratio of capacity to weight or volume is required. Batteries used for private motor-cars, aircraft and light vans have thin pasted plates about 1·8 to 2·5 mm (0·07 to 0·1 in.) thick. The close and compact assembly of thin plates results in a battery of low internal resistance and therefore minimum voltage drop when delivering large currents for engine starting. The largest and thickest plates, up to about 6·4 mm (0·25 in.) thick, are used for heavy traction and commercial vehicle batteries, where a long life under arduous operating conditions is required. One important type of positive plate for heavy traction work is of tubular design used in conjunction with a flat pasted negative plate, but usually in portable batteries both positive and negative plates are of the flat pasted type.

GRIDS

The grid is the framework, or lattice, which supports the paste, or active materials. Grids for positive and negative plates are of the same basic design, although the negative grid is usually somewhat thinner as it is not subjected to the same corrosive wear and tear in service as the positive grid.

Various types of grid are shown in Fig. 2.1. The grid is made mainly of antimonial-lead alloy, consisting of pure lead to which has been added from 6 to 12 per cent of antimony. Frequently small quantities of other metals such as tin, copper, silver or arsenic are added to give greater corrosion resistance or similar properties.

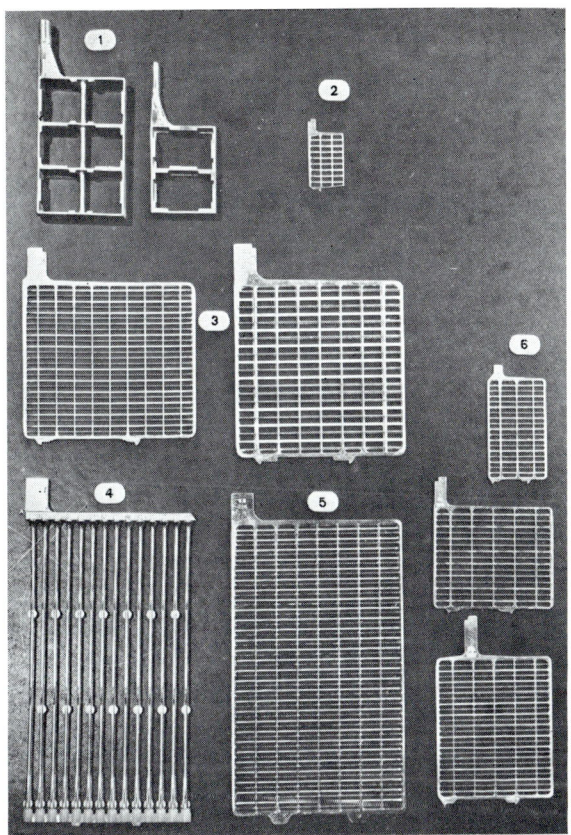

FIG. 2.1. VARIOUS TYPES OF GRID

1. Heavy grids for "low-loss" cells   4. Grid for tubular positive plate
2. Miniature grid   5. Grid for negative plate of tubular cell
3. Automotive battery grids   6. Scooter and motor-cycle battery grids

The addition of antimony to the pure lead produces a sharper and stronger casting than that of lead alone, and also gives it a greater resistance to the electrolytic action and chemical changes which take place in the active material which the grid supports.

A grid consists of an outer frame with take-off lug and a central mesh or lattice of vertical and horizontal ribs. The ribs serve as conductors for collecting and distributing the current throughout the active material, and as retainers for the paste. Fig. 2.2 (*b*) is a cross-section of the grid showing the staggered horizontal ribs, which are designed to provide a continuous ribbon of paste from top to bottom of the plate.

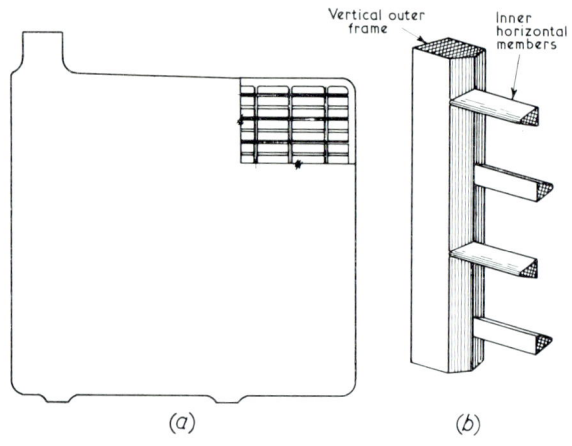

(*a*)   (*b*)

FIG. 2.2. CAR BATTERY PLATE
(*a*) Cut away to show portion of grid
(*b*) Enlarged cross-section of grid

PLATES

Pasted plates are made by applying to the grid, by hand or by machine, a paste consisting mainly of lead oxides and dilute sulphuric acid. The pasted plates are dried and then *formed* electrochemically in a tank containing dilute sulphuric acid. Formation of pasted plates produces oxidation of the lead oxide paste to lead dioxide (positive) and its reduction to spongy lead (negative). When formation is completed the plates are taken out of the acid tanks, rinsed in water and dried. In the tubular plate construction, dry lead oxide is shaken into the tubes threaded over the grid spines. The open end is then sealed by a polythene bar.

POSITIVE AND NEGATIVE GROUPS

The dry, formed plates suitably spaced apart are assembled into positive and negative groups by burning the plate lugs to a plate bar or strap complete with terminal post (Fig. 2.3).

A negative group always has one more plate than its matching positive group, so that when the two groups are interleaved each positive plate is located between two negative plates. This arrange-ment ensures that the surfaces of each positive plate are worked equally, and prevents distortion or buckling of the positive plates, which would occur with unequal working of their active material.

ELEMENTS

When positive and negative groups are interleaved, adjacent plates of opposite polarity must be prevented from touching each other;

FIG. 2.3. NINE-PLATE ELEMENT, 4-PLATE POSITIVE GROUP AND 5-PLATE NEGATIVE GROUP

otherwise a short-circuit would develop within the cell. Separators of various materials are inserted between the plates, and the assembly of plate groups and separators is called an *element*. A 9-plate element would consist of 4 positive plates, 5 negative plates and 8 separators (Fig. 2.3).

SEPARATORS

The essential features of any separator are—

1. High porosity; this ensures low resistance to passage of current between the plates and free diffusion of the acid.
2. Good insulation, to prevent metallic conduction between plates of opposite polarity.
3. It must be inert to the action of sulphuric acid and electro-chemical oxidation.
4. Absence of harmful impurities.

5. Mechanical strength–separators which are easily damaged or split are a source of internal short-circuits.

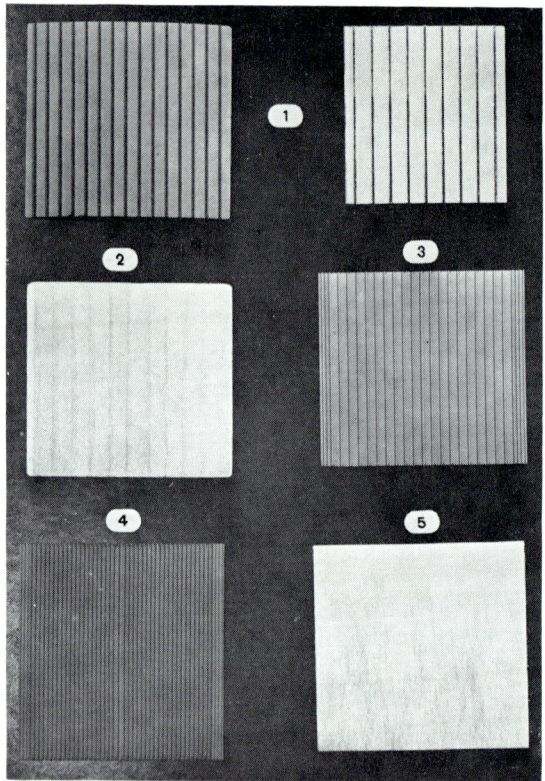

FIG. 2.4. VARIOUS SEPARATORS FOR AUTOMOTIVE BATTERIES

1. Fibre-based resin-treated
2. Glass-wool/kieselguhr
3. Porvic, microporous p.v.c.
4. Porvic, microporous p.v.c. used with
5. Glass-wool mat for dual separation of heavy-duty batteries

Various types of separator are shown in Fig. 2.4. These may consist of any suitable material such as—

1. Microporous rubber.
2. Microporous plastic.
3. Paper base (resin treated).
4. Certain kinds of wood, the most suitable being Port Orford cedar or Douglas fir.

5. Glass-wool mats.
6. Glass-wool and kieselguhr.

Before 1940 most separators used in portable batteries were of wood, but since about 1946 these have been replaced mainly by microporous plastic (polyvinyl chloride) or microporous rubber, which are highly suitable for this type of work.

The width and height of separators are always greater than those of the plates. This provides a separator overlap at the plate edges and so prevents lead *treeing*, or formation of lead moss, across adjacent plates. Most separators are ribbed or grooved on the side adjacent to the positive plate, so as to ensure a liberal supply of free electrolyte, which is essential for efficient working of the positive active material. Cells fitted with tubular positive plates use flat sheets of porous plastic or rubber separators, as the tubular construction ensures an adequate space for acid.

Batteries intended for heavy duty as in traction service (road vehicles, industrial trucks) or in heavy commercial vehicles (buses, etc.) use a dual separation of microporous rubber or microporous plastic separator, together with a mat of glass-wool. In this assembly the plates and separators are compacted tightly together with the glass-wool mat adjacent to the surface of the positive plate. The glass-wool mat serves as an "armouring" for the positive plate by supporting and retaining the active material in the grid. This extends battery life, particularly on heavy work, by preventing shedding of the positive paste from the grid.

CONTAINERS

In general, the containers used for portable batteries are made of hard rubber, resin rubber or composition material. More use is being made of polystyrene and polythene containers, but these are generally confined to small batteries used for motor cycles, scooters, radio and light-weight aircraft batteries. Containers are either individual boxes for a single element or multi-compartment moulded containers for housing 2 to 6 elements (Fig. 2.5). A 12-V battery would consist of 6 elements housed in a 6-compartment (monobloc) container. Each cell must be a separate electrical unit, consisting of element and electrolyte, insulated from the next cell by the compartment partitions.

The plates of most portable cells are never allowed to sit on the bottom of the box but rest on ribs or grids moulded in the base. This arrangement provides a space below the plates for sediment, or mud. Sediment is an accumulation consisting largely of lead sulphate which is thrown down as active material from the plates during life, particularly if the battery is subjected to repeated cycles of heavy discharge and charge, or abused by excessive overcharging or undercharging. As the sediment acts as a metallic conductor, the space below the plates must be sufficiently deep to keep the sediment away from the bottom of the plates during the life of the battery.

FIG. 2.5. HARD-RUBBER MOULDED. SINGLE-CELL BOX AND 6-CELL MONOBLOC CONTAINER

Both have been cut away to show the base supporting ribs

Large cells, as used in traction or marine batteries, are housed in separate moulded or wrapped hard-rubber boxes for convenience in handling.*

LIDS

Battery lids are moulded in hard rubber or composition material. They are made as single units with three holes, two for the terminal posts, and the third for the vent plug. Alternatively, the lid can be made as one piece, or "monolid" (Fig. 2.6). In either form the lid is fitted over the cell posts, and an acid-tight seal is made between lead

---

* *Wrapped box*. A box prepared by wrapping thin layers of soft processed rubber round a metal former. A tough, rigid box is produced by curing or baking in a steam oven.

post and lid, either by pressing against a rubber gasket or by lead-burning the cell post to a lead insert in the lid.

The cell is acid-proof sealed by pouring a sealing compound in the trough between the lid and the container walls.

Vent Plugs

The vent plugs fit in the holes of the lid which are provided for filling the battery with electrolyte, for topping-up with distilled

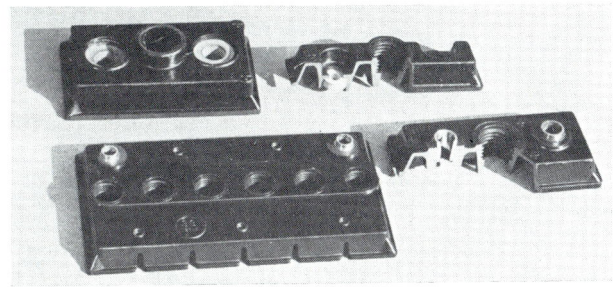

FIG. 2.6. Various Lids

The two lids in the upper part of the picture are designed for rubber gasket seal; the two in the lower part are fitted with lead bushes (lid-insert type of seal)

water or taking hydrometer readings. The vent plugs are of the baffle type which allow the gases to escape freely but return the acid spray to the cell. Modern vent plugs are moulded in polystyrene and are lighter in weight than the older ones of rubber or porcelain. They are either screw or quarter-turn type, the latter being used in multicell traction batteries for ease of removal.

Connectors

Connectors are antimonial lead or lead-plated copper intercell straps burned or bolted to adjoining positive and negative posts of adjacent cells or elements (Fig. 2.7).

The type of connector most commonly used for automotive batteries is made from a casting of antimonial-lead alloy. Lead-plated copper connectors have the advantage of greater flexibility and are used as intercell connectors for most traction batteries. Connectors are designed to be of low resistance and capable of carrying a current in amperes equal to about $5C$, where C is the nominal ampere-hour capacity, without any undue voltage drop or heating.

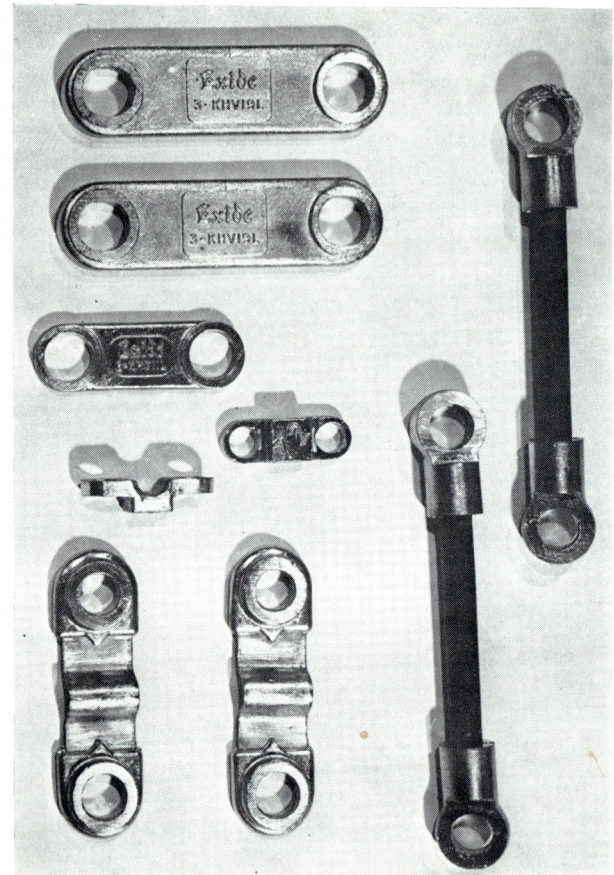

FIG. 2.7. VARIOUS CONNECTORS FOR AUTOMOTIVE AND TRACTION
BATTERIES

ELECTROLYTE

A battery when constructed must be filled with dilute sulphuric acid and given an initial charge to activate the plates. The acid must be chemically pure as specified in B.S. 3031:1958.

Stationary Battery Construction

From the very earliest days the Planté-type battery has been used for most stationary battery applications. The Planté plates first used were about 12 mm (0·5 in.) thick, and because of their very robust character were ideally suited for heavy work involving cycles of discharge and charge, and for installations where weight and space imposed no limitations on battery size. Modern stationary batteries are used mostly for emergency or standby duties, which means that generally the batteries are smaller both in physical size and electrical storage capacity. Where floor space and structural strength of existing buildings are critical factors, batteries using pasted plates are also used for stationary duties. These do not have quite the same long life as the Planté battery, but a compromise has been achieved by the introduction in the last few years of a high-capacity Planté plate 8 mm (0·31 in.) thick. Planté cells assembled with this new plate give almost $2\frac{1}{2}$ times the output of the old heavy Planté cell in the same volume, or occupy considerably less space for the same capacity.

PLANTÉ CELL CONSTRUCTION

The Planté plate is a casting of pure lead consisting of numerous thin vertical laminations, strengthened by a series of horizontal

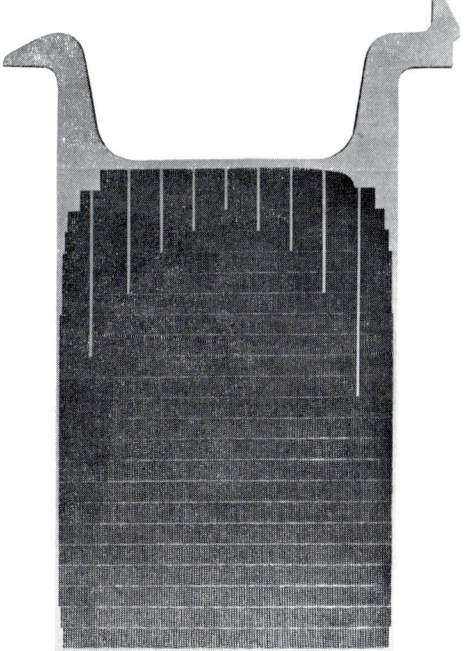

FIG. 2.8. PLANTÉ POSITIVE PLATE

cross-ribs (Fig. 2.8). The effect of the laminations is to increase the superficial area of the plate by as much as 12 times that of a plain lead plate of similar width and length. Almost the whole of this

FIG. 2.9. ROSETTE POSITIVE PLATE

developed area is available for conversion to active material (lead dioxide) by electrochemical formation of the pure lead surface.

The earliest Planté plates were formed by a series of cycles of charge and discharge in dilute sulphuric acid. This was a very lengthy and tedious procedure; modern formation processes are very much quicker. A forming agent is added to the sulphuric acid, which attacks the surface of the lead laminations and accelerates the production of the lead dioxide coating when current is passed through the plate during formation.

At one time the Planté cell used both Planté-type positive and negative plates, and the two sets were formed at the same time, one set at the end of the process being lead dioxide (positive), and the other spongy lead (negative). However, Planté negatives have been obsolete for many years, and it is modern practice to form Planté positives against plain lead sheets, or "dummies."

A second type of Planté plate, which is still widely used in America but is almost obsolete in this country, is the rosette or Manchester plate. This is an antimonial-lead casting perforated with numerous holes in which are plugged rolls of previously crimped lead tape, or "rosettes" (Fig. 2.9). During formation the surface of the lead tape of the rosettes is converted to the lead dioxide active material.

FIG. 2.10. BOX NEGATIVE PLATE, UNPASTED
The lead frame is cast in two halves

NEGATIVES

Two types of negative plate are used with the Planté positive: a box negative with the heavy Planté plate, and a pasted negative with the lighter high-performance Planté plate.

The box negative consists of a pair of antimonial-lead alloy grids, each with an outer surface covered with a thin sheet of perforated lead. After applying the paste the two grids are riveted together (Fig. 2.10). The pasted negative plate is merely a much larger version of the type used for portable batteries.

Cell Construction

A study of Figs. 2.11 and 2.12, showing exploded views of the two

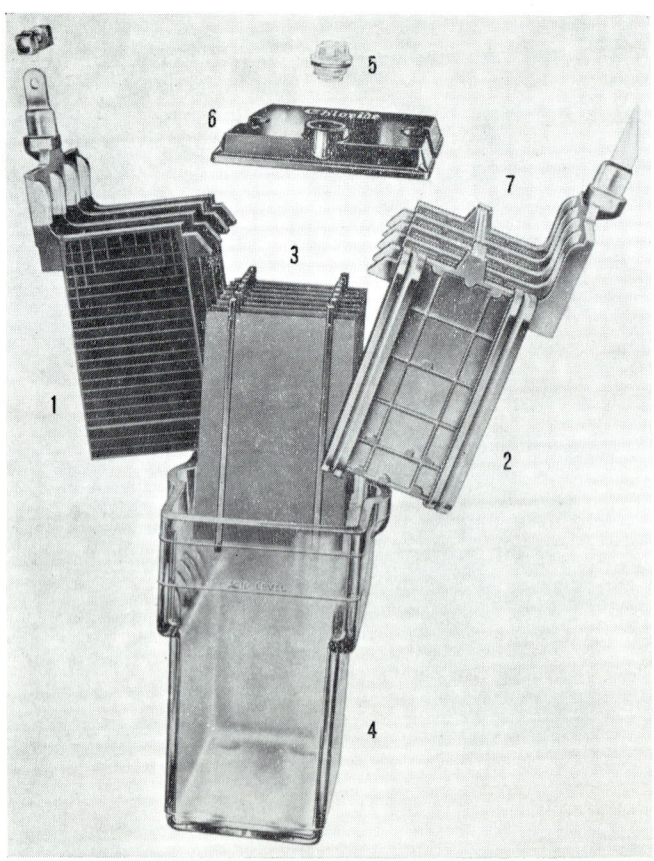

FIG. 2.11.   EXPLODED VIEW OF PLANTÉ CELL, ENCLOSED TYPE IN MOULDED GLASS BOX

1. Positive plate group
2. Negative plate group
3. Separator diaphragms and dowels
4. Glass box

5. Vent plug
6. Hard-rubber lid
7. Separator hold-down rod

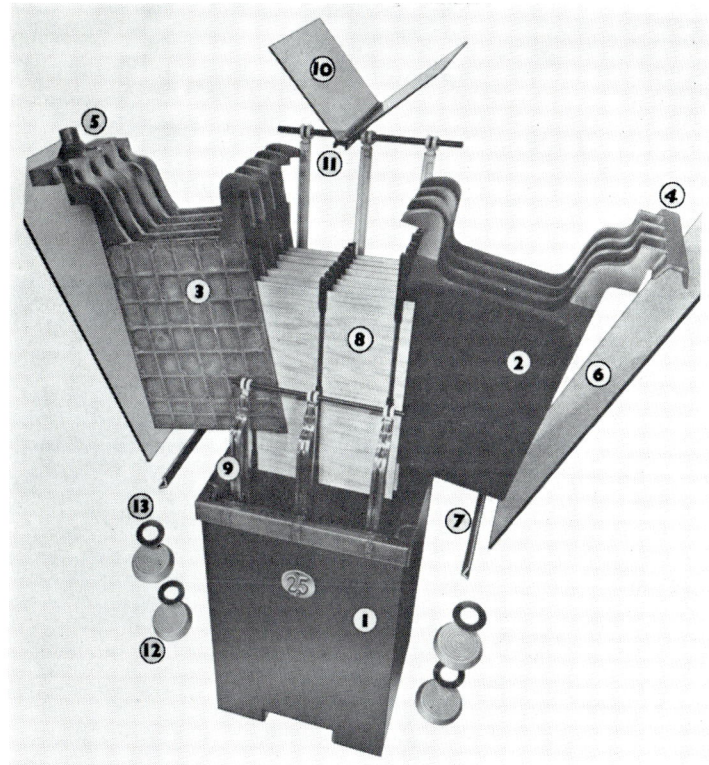

FIG. 2.12. EXPLODED VIEW OF OPEN-TYPE PLANTÉ CELL, IN WOOD
(LEAD-LINED) BOX

1. Lead-lined wood box
2. Positive plates
3. Negative plates
4. Channel bar
5. End bar with cup
6. Glass hangers
7. Lead supports for hangers
8. Separators and dowels
9. Buffers and supporting tubes
10. Spray arrestor assembly
11. End tube assembly
12. Insulators
13. Packing discs

main types of stationary cell, will be helpful in understanding the
component parts of the cell detailed below.

PLATE GROUPS

Positive and negative groups are assembled by burning the individual plates to the lead terminal bars. As in the pasted plate
assembly, in any two matched groups there is always one more
negative plate than positive, so that when they are interleaved the
end plates are always negative. This arrangement is more essential

in Planté cells than in pasted cells because of the necessity to work both surfaces of the Planté positive plate. During the life of the cell, lead dioxide is gradually lost from the plate laminations, and is made good by conversion of the underlying surface to lead dioxide during the normal charging and discharging of the cell. It is this characteristic which gives the Planté positive its very long life.

SEPARATORS

The separators may consist of wood veneers, microporous plastic veneers or glass tubes.

Glass tubes of different length and diameter to suit the various cell sizes are pushed down between the plates so as to rest on the bottom of the cell box. They are held in a vertical position by location in guides in the top of the plates.

Wood separators are made of thin wood sheets threaded through two or more slitted wooden rods or dowels which support the veneers in the correct position between the plates. Wood separators are now largely being replaced by sheets of microporous plastic threaded through slitted rods of polystyrene.

CONTAINERS

Plate groups are assembled in glass boxes sealed with lids, open glass boxes, or lead-lined wood boxes. Glass is used for containers for cell capacities up to about 2,000 Ah. Cells of larger capacity, up to an existing maximum of 15,050 Ah, are assembled in lead-lined wood containers. All containers are made much deeper than the length of plate they accommodate so as to allow ample gassing space above the plates, and space below for the downward expansion of the positive plates and the build-up of sediment.

When a glass box is used, the plate groups are supported by the plate lugs resting on the top of the box or on shoulders moulded in the box. In lead-lined wood boxes, the lining is carried over the top edges of the box, and the plate lugs are supported on glass slabs projecting above the top edges. These glass slabs are located in lead grooves in the base of the container.

Cells may be interconnected either by bolting the terminal bars together or burning the plate lugs of adjacent cells to a common terminal bar.

In recent years cells with positive plates of tubular construction have been used for stationary applications. Their construction is very similar to that used for portable traction applications, except that in some instances the hard-rubber box is replaced by a glass box.

Mass-type Cell

The mass-type cell is a special cell which is used for both portable and stationary applications. It was originally designed for supplying low-tension current in the early radio receivers, and since then it has been used for all kinds of low-current duties such as laboratory work, electric clocks, fire alarms, electric fences, electric bells, and small telephone installations. A typical cell is shown in Fig. 1.1, page 6, where it will be observed that there are only two plates: one positive, about 10 mm (0·4 in.) thick; and one negative, about 15 mm (0·6 in.) thick, assembled, without any separator between the plates, in a plastic container.

The Grouping of Plates, Cells and Batteries

Plates for storage batteries vary considerably in size and thickness, the largest plate (Planté) measuring about 51 cm by 76 cm by 13 mm thick, and weighing 36·3 kg, compared to the smallest plate (pasted) measuring 30 mm by 44 mm by 1·2 mm thick, and weighing 15 g. The largest plate has a capacity of 432 Ah at the 10-hr rate of discharge, compared with 0·5 Ah for the smallest pasted plate. The largest cell made is a Planté type, having a capacity of just over 15,000 Ah, weighing over 3000 kg, and measuring 130 cm high by 170 cm long by 70 cm wide.

The width and height of the plates determine the width and height of the box in which they are assembled; the length of the box is variable, depending on the capacity required. For example, using positive plates each of 100 Ah capacity, and requiring a cell of 1,000 Ah capacity, would mean using 10 positive and 11 negative plates to complete a 21-plate cell. Similarly a 2,000-Ah cell using the same plates would require 20 positives and 21 negatives, and the length of the box would be almost twice that of the 1,000-Ah cell. Alternatively, a cell of 2,000 Ah could be assembled using 10 positives each of 200-Ah capacity in the same box length as the 1,000-Ah cell, but with increased width and height to accommodate the larger plates.

The largest portable cell in normal use is made for traction purposes, measuring 74 cm high by 30 cm long by 23 cm wide, with a weight of 155 kg and a capacity of 2,200 Ah.

Cells for special portable duties such as submarine applications are larger; one of these has a capacity of 8,000 Ah and weighs nearly 550 kg.

Although lead-acid cells vary considerably in ampere-hour capacity and dimensions, from the size of a match box to that of a large domestic refrigerator, the voltage of any cell is approximately 2 V, irrespective of its size.

GROUPING OF CELLS AND BATTERIES

A storage battery consists of two or more cells connected in series, and as shown in Fig. 2.13, the diagrammatic way of representing them is by a long thin line for the positive pole of the cell, and a short thick line for the negative pole. Pairs of strokes may be used to indicate the number of cells in a battery—3 pairs for a 6-V lead-

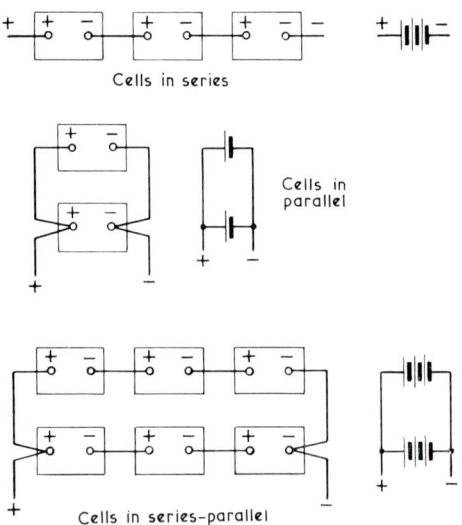

FIG. 2.13. DIAGRAMMATIC REPRESENTATION OF CELLS AND GROUPING

acid battery, 6 pairs for a 12-V battery—but for batteries of more than 6 cells, usually the first and last pairs only are shown, joined by a dotted line.

For alkaline batteries there are 5 cells to a 6-V battery, and 9 cells to a 12-V battery, because the alkaline cell has a nominal voltage of only 1·2 V.

Cells are said to be in *series* when the positive pole of one cell is connected to the negative pole of the adjacent one, and this arrangement is continued for any desired number of cells. The greater the number of cells in series, the greater will be the voltage of the battery, as the voltages of the cells are additive. The capacity of the battery will, however, still be that of a single cell.

Cells are connected in *parallel* when all the positive poles are joined together and all the negative poles are similarly connected. (Fig. 2.13). The voltage of cells connected in parallel is that of a single cell, but the capacity of the combination is the sum of the individual cell capacities.

Series-parallel arrangements of batteries, or cells, are usually made for convenience when charging a large number of batteries at the same time from a limited-voltage d.c. supply.

When connecting two or more batteries in parallel, all must have the same voltage, that is, the same number of cells in series. Batteries of the same voltage but of different capacity may be connected in parallel both for discharge and charge, as they will take a share of the total current proportionate to their respective capacities.

Equilibrium conditions will exist as long as the cells in each parallel branch are maintained in the same healthy condition. A weak or defective cell in one of the batteries means that this battery on discharge does not carry its share of the total load. On charge it accepts too great a share of the current available, to the detriment of the other batteries in parallel with it.

CHAPTER 3

GENERAL ELECTRICAL CHARACTERISTICS

In selecting a storage battery as a source of electrical power for a particular duty, several factors have to be considered. The nature of the application or duty, that is, whether stationary or portable, influences the size and weight, the type of plate and container of the cell or battery to be offered. The electrical specification relating to the voltage of the system determines the number of cells in the battery, and the time during which current must be supplied determines its capacity.

Capacity of Storage Batteries

The *capacity* of a storage cell, or battery, can be expressed either in ampere-hours or in watt-hours. Thus one of the units of capacity is the *ampere-hour* (Ah), capacity in this case being the product of the current in amperes (A) and the time in hours (h).

Type of battery	Nominal capacity rating	Specified temperature		Specified final voltage, per cell
	hr	deg C	deg F	V
Lead-Acid				
Automotive . .	20	25	77	1·75
Diesel starting . .	5	21	70	1·73
Traction (truck and vehicle) . . .	5	27	80	1·70
Planté and rosette .	10	15·6	60	1·85
Train lighting and air-conditioning . .	10	15·6	60	1·80
Alkaline . . .	5	20	68	1·10
		or		
		27	80	

A battery with a rated capacity of 100 Ah at the 10-hr rate can supply 10 A for 10 hr. At rates of discharge in excess of 10 A the battery will provide less than 100 Ah, whilst at rates less than 10 A the battery will provide more than 100 Ah.

Thus, in specifying the capacity of a battery it is necessary to state the time rate of discharge. For example, a heavy-plate Planté battery giving 100 Ah at the 10-hr rate would give only 50 Ah at the 1-hr rate. The capacity of a battery also varies with its temperature and final voltage, so all these values must be stated. In the case of this 100-Ah battery, its capacity would be related to the 10-hr rate at a temperature of 15·6° C (60° F), and to a final voltage of 1·85 V per cell (p.c.). This could be written: 100 Ah/10 hr/1·85 V p.c./ 15·6° C. An automotive battery of 100-Ah capacity would be related to the 20-hr rate at 25° C (77° F), and to a final voltage of 1·75 V per cell (100 Ah/20 hr/1·75 V p.c./25° C). A traction battery of 100-Ah capacity would be related to the 5-hr rate at 27° C (80° F), and to 1·70 V per cell (100 Ah/5 hr/1·70 V p.c./27° C).

The table on page 30 gives the conventional standards usually adopted in this country for nominal capacity ratings, temperatures and final voltages of the various types of battery.

WATT-HOUR CAPACITY

When discharged at the nominal rates shown in the table, lead-acid batteries will commence discharge at a voltage of approximately 2·0 V per cell, and alkaline cells at approximately 1·35 V per cell. The average voltage during discharge will be some value between the initial and final voltages, and will vary according to the specified final voltage. A Planté cell at the 10-hr rate to 1·85 V final voltage will have a higher average discharge voltage than a traction cell with a final voltage of 1·70 V.

The product of the average discharge voltage and the ampere-hour capacity gives the *watt-hour capacity* of a battery.

Consider a Planté and a traction battery of the same nominal ampere-hour capacity of 100 Ah—

The Planté battery would be quoted as having a rating of

100 Ah/10 hr/1·85 V p.c./15·6° C

At the 10-hr rate the average discharge voltage would be 1·94 V per cell, and the watt-hour rating would be

100 × 1·94, or 194 Wh per cell

The traction battery would be quoted as having a rating of

100 Ah/5 hr/1·70 V p.c./27° C

At the 5-hr rate the average discharge voltage would be 1·91 V per cell, and the watt-hour rating would be

100 × 1·91, or 191 Wh per cell

The watt-hour capacity of the traction battery of N cells would be $100 \times N \times 1\cdot91$, or $191 \times N$ watt-hours.

For nominal watt-hour ratings it is usual to take 2 V per cell as the average discharge voltage.

PERCENTAGE-CAPACITY/TIME CURVES

The capacity of a lead-acid battery varies with the rate of discharge, and the capacity available from a battery, whether in ampere-hours or in watt-hours, is greatest at low rates and lowest at high rates of discharge.

The relationship between capacity and time of discharge is shown in Fig. 3.1 for Planté, automotive, and traction batteries. Reading from the curves it is possible to estimate the capacities of the different batteries at various rates of discharge—

Capacity	Planté	Automotive	Traction
Nominal	100 % (10 hr)	100 % (20 hr)	100 % (5 hr)
10-hr	100 %	89 %	111 %
5-hr	83–88 %	78 %	100 %
3-hr	72–79 %	70 %	89 %

The reduction of capacity at high rates of discharge is due to the rapid formation of lead sulphate at the surface of the plates. This blocks the pores of the plates, retarding and finally preventing further diffusion of the acid to the inside active material. The rapid formation of sulphate also increases the internal resistance of the battery, which in turn reduces the voltage available at its terminals.

At slow rates of discharge the diffusion of the acid into the pores of the active material proceeds slowly and sulphate is not formed so quickly at the plate surface. More of the active material is available for conversion to lead sulphate, more ampere-hours are obtained, and the battery voltage remains at a steady value for a longer period.

Although a battery may be discharged at a high current to the stage where its voltage falls off rapidly, it will, if allowed to stand on open-circuit for a short time, recover in voltage and be capable of delivering further current. This phenomenon is referred to as *recuperative capacity*.

Similarly, following a high current discharge to the stage of collapse, a battery will recover and be capable of delivering considerable capacity if the current is reduced.

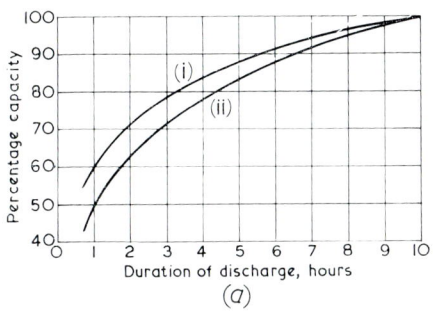

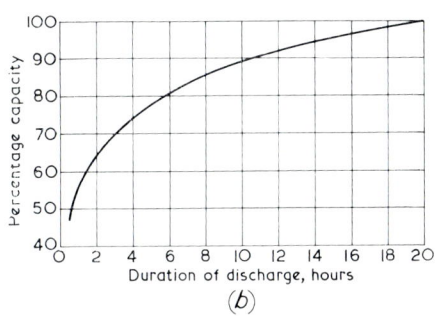

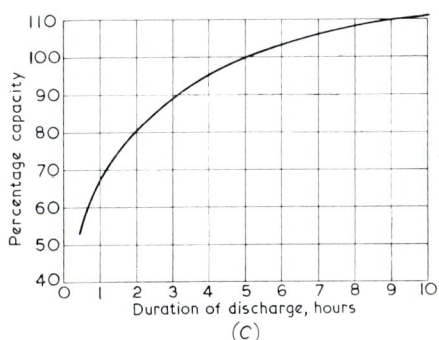

Fig. 3.1. Ampere-hour-Capacity/Time Curves

(a) Planté batteries
  (i) High-performance plates
  (ii) Heavy plates
(b) Automotive batteries
(c) Traction batteries

Factors affecting Capacity

There are many factors which directly influence the capacity of lead-acid batteries, the most important being

1. Rate of discharge
2. Practical limit of final voltage
3. Temperature of battery
4. Amount of active material
5. Design and number of plates
6. Volume and specific gravity of electrolyte
7. Age of battery

Final Voltage

So far, the effects of the rate of discharge on the available capacity of a battery have been considered. The capacity can be related to any practical limit of voltage, and the higher the final voltage the

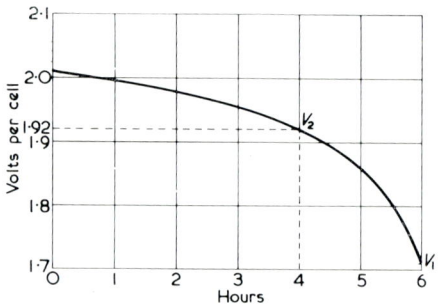

FIG. 3.2. EFFECT OF FINAL VOLTAGE ON CELL OUTPUT

With load equal to 6-hr rate, nominal final voltage is V_1
Duration to permissible voltage, V_2, is 4 hr

lower will be the capacity available. This can be demonstrated for a typical duty where a 24-cell battery supplies a load to a minimum working voltage of 46 V, or 1·92 V per cell. The current drawn from the battery during peak loading might be the equivalent of the 6-hr rate to the standard final voltage of 1·71 V per cell, but when related to the permissible working voltage of 1·92 V per cell, the battery would supply the current for only 4 hr (Fig. 3.2).

The final voltage is determined partly by the nature of the duty which the battery performs and partly by the discharge voltage characteristic of the battery. A lighting load requires a fairly high stable voltage, whilst a heavy engine-starting load requires a high power output from the battery, that is, high current at a lower working voltage and to a lower final voltage.

When final voltage is determined by the discharge voltage characteristic of the battery, it is usually fixed to allow a fall of 0·3 V per cell from the initial voltage. The cell voltage on discharge falls sharply beyond this end point, which is known as the *knee* of the voltage curve. Continuing the discharge beyond the knee does not produce any worth-while increase in available capacity, because of the sudden collapse of cell voltage.

Temperature

Normal temperature for storage batteries is generally accepted as falling within the range from 15·6° to 27° C (60° to 80° F). Low battery temperatures temporarily reduce the available ampere-hour capacity and discharge voltage. Capacity and voltage are restored with a return to normal temperature, even without a charge. An increase in battery temperature results in an increase in capacity, particularly at high rates of discharge.

Both these effects of temperature on capacity are mainly a result of the change in the viscosity and resistance of the electrolyte. At low temperatures the viscosity and resistance are increased, especially below 0° C (32° F). The increased viscosity reduces the rate of diffusion or circulation of electrolyte into the pores of the active materials. At high temperatures the viscosity and resistance are reduced and the battery capacity is increased.

The effect on battery capacity of changes in the viscosity and resistance of the electrolyte are most noticeable at high rates of discharge, where electrolyte diffusion is the main factor in limiting the available capacity.

The standard temperatures for the different types of battery are shown in the table on page 30.

Stationary battery capacities are rated for 15·6° C (60° F) or 25° C (77° F).

Portable battery capacities are rated for approximately 21° C (70° F) or 25° C (77° F).

Traction battery capacities are rated for 27° C (80° F).

These are approximately the average operating temperatures of the different types of battery. The temperature of the traction battery is the highest because this type is usually the hardest worked, doing a regular daily cycle of almost complete discharge and recharge. This produces an appreciable amount of heat within the battery.

TEMPERATURE CORRECTION FACTOR

When batteries are used in cold climates where the electrolyte temperature falls to $0°$ C ($32°$ F) or lower, allowance must be made for the reduction in available capacity.

A traction battery of 100-Ah capacity at the 5-hr rate at $27°$ C ($80°$ F) would give 85 Ah at $0°$ C ($32°$ F), equivalent to a reduction in capacity of 0·56 per cent per deg C.

A Planté battery of 100-Ah capacity at the 10-hr rate at $15·6°$ C ($60°$ F) would give 86 Ah at $0°$ C ($32°$ F), equivalent to a reduction in capacity of 0·56 per cent per deg C.

An automotive battery of 50-Ah capacity at the 20-hr rate at $25°$ C ($77°$ F) would give 38 Ah at $0°$ C ($32°$ F), equivalent to a reduction in capacity of 0·96 per cent per deg C.

The temperature correction factor for capacity increases with fall in temperature and increase in rate of discharge.

The automotive battery at the 5-min rate would have a temperature correction factor of 1·3 per cent per deg C. This means that at $25°$ C the 50-Ah battery would supply 168 A for 5 min, whilst at $0°$ C it would supply only 115 A.

Amount of Active Material

The electrical energy that a battery delivers during discharge is derived from the electrochemical reactions taking place between the electrolyte and the active materials of lead dioxide and spongy lead. The greater the amounts of these materials, the greater will be the capacity of the battery.

The theoretical relationship between the weight of active materials and electrical output in ampere-hours of a storage cell can be calculated by reference to Faraday's electrochemical law. The theoretical amounts of active material required for 1 Ah of electricity are 4·46 g of lead dioxide and 3·87 g of lead. In practice, from three to five times the theoretical amounts are required, depending on the type of cell and the thickness and number of plates.

Design of Cell

For a cell to give the maximum output for a given weight of active material, it is essential for the active material to be readily accessible to the electrolyte. This is achieved by using a large number of thin plates, with a high weight ratio of active material to grid, compatible with maintaining the grid sufficiently strong for good conductivity and retention of paste. The same weight of active material distributed in a few thick plates containing a dense or hard paste would result in a cell of much lower capacity.

A plate with porous active material is capable of producing a greater output, particularly at high rates of discharge, than a plate with a dense, hard, active material. Porosity is produced by adding small amounts of carbon, barium sulphate or other organic or inorganic expanders during the mixing of the paste. Too high a porosity is avoided for plates which are to be subjected regularly to cycles of discharge and charge, as this is conducive to short life by premature shedding of the active material.

The formation of lead sulphate during the normal discharge of a cell reduces the porosity of the plate, because the lead sulphate occupies more space than the original active material. Normally this expansion is taken up by the pores of the plates being compressed, and no harm results. If, however, the sulphation is excessive, owing to the cell being persistently overdischarged and undercharged, or left for long periods in a discharged condition, the sulphate expands beyond the absorbing capability of the pores. The active material is then subjected to pressure which results eventually in fracture of the grid frame or loss of active material from the plate, and if the condition is not remedied the cell loses capacity and will not accept a charge.

INTERNAL RESISTANCE

When the terminals of a storage cell are joined to an external circuit of resistance R ohms, a current of I amperes will flow which can be measured by connecting an ammeter in the circuit. This current also flows inside the cell, being conducted from the positive to the negative plates by the electrolyte. The total resistance is therefore that of the external circuit, plus the internal resistance of the cell. By Ohm's law,

$$I = \frac{E}{R_E + R_C}$$

where E is the electromotive force, or open-circuit voltage, of the cell, R_E the resistance of the external circuit, and R_C the internal resistance of the cell.

The internal resistance of a storage cell varies with temperature, state of discharge and design of cell. It increases with fall in temperature and extent of discharge, and decreases during charge. The internal resistance of any storage battery is low compared with the resistance of any external circuit to which it is likely to be connected.

If the above equation is rearranged as

$$E = IR_E + IR_C$$

it will be appreciated that, when R_E is high and I small, the voltage drop IR_C within the cell itself will be negligible compared to IR_E. When I is a hundred or more amperes, as in engine-starting or similar applications, IR_C may be as high as 0·1 V or more. At these heavy currents an appreciable part of the energy of the cell is wasted in internal heating, which is proportional to the square of the current (I^2R_C).

It is obvious that when the cell or battery is intended to provide heavy currents the internal resistance must be kept low. This is achieved by designing a cell with many thin plates with a short path of electrolyte between them. Plates forming the same group are in fact in parallel, so that the effective resistance of N plates, each of resistance R_P, is $R_P \div N$.

It is sometimes convenient to express the capacity of a cell in ampere-hours per positive plate. A cell of 9 plates has 4 positive and 5 negative plates. Assuming that each positive plate has a capacity of 10 Ah, the capacity of the cell will be 4 × 10 = 40 Ah. Using the same size of plate, a cell of 21 plates has 10 positive plates and a capacity of 10 × 10 = 100 Ah.

Similarly, it is convenient to specify internal resistance in ohms per positive plate. The 10-Ah plate may have a resistance of 0·01 Ω, and the 9-plate cell has 4 positive plates in parallel, giving a resistance of 0·01 ÷ 4, or 0·0025 Ω.

A battery of six 9-plate cells would have a nominal voltage of 12 V (6 × 2), a capacity of 40 Ah, and an internal resistance of 6 × 0·0025, or 0·015 Ω, equivalent to the resistance of 6 cells in series.

The total resistance of the battery would be the internal resistance plus the resistance of 5 inter-cell connectors. The resistance of the connectors would amount to approximately one-sixth of the internal resistance of the battery. The total resistance of the battery would therefore be 0·015 + 0·0025 = 0·0175 Ω.

The internal resistance of a battery increases with extent of discharge, and the fully discharged resistance is $2\frac{1}{2}$ to 3 times the fully charged value.

Electrolyte

The electrolyte used in lead-acid batteries is a solution of pure sulphuric acid (H_2SO_4) in pure water. The specification for purity of the sulphuric acid to be used is given in B.S. 3031:1958.

In service, the electrolyte can be contaminated by using impure water for replacing that lost by gassing and evaporation, the general effect being to reduce the life and performance of the plates. This will be considered in greater detail in a later chapter.

VOLUME AND DENSITY

In an earlier chapter the chemical reactions taking place during the discharge and charge of lead-acid cells were considered. It was shown that the electrolyte plays an equally important part in the working of the cell as do the other active materials, and without a sufficient volume of electrolyte the cell cannot work efficiently.

Previously, reference has been made to the theoretical amounts of active material required to produce 1 Ah, namely 4·46 g of lead dioxide and 3·87 g of spongy lead. The theoretical amount of sulphuric acid required to take part in the electrochemical reactions is 3·66 g.

In practice, the actual amount of electrolyte required to produce 1 Ah exceeds the theoretical amount, just as the theoretical amounts of active materials are also exceeded. All the acid in a cell is not utilized in the reactions, and this is particularly true of the acid below the plates, and to a lesser degree of the acid at the sides of the element and above the tops of the plates.

In most stationary-type cells, where the plates are widely spaced, there is an ample volume of electrolyte for the capacity required. In such cells the density or specific gravity (sp. gr.) of the electrolyte can be lower than that used in the more compact portable types of cell, where plates are packed together with little acid space between adjacent plates.

SPECIFIC GRAVITY RANGE

The specific gravity of the electrolyte for the main types of lead-acid cell at 15·6° C (60° F) varies as shown—

Stationary cells .	.	.	.	.	.	1·210
Portable cells—						
Temperate climate .	.	.	.	.	.	1·270–1·285
Tropical climate	.	.	.	.	.	1·230–1·250

The specific gravity falls during discharge and increases during charge. It is least when the cell is fully discharged, and greatest when the cell is fully charged.

The specific gravity reading therefore provides an excellent indication of the state of discharge of a lead-acid battery, as the fall in specific gravity during discharge is directly proportional to the number of ampere-hours supplied by the battery.

The difference between the values of specific gravity with the cell fully charged and fully discharged varies according to the amount of electrolyte in the cell. The usual working ranges of specific gravity for the main types of cell are approximately

Stationary .   . 1·210 to 1·130–1·180 (30 to 80 points) at the 10-hr rate

Automotive .   . 1·280 to 1·110 (170 points) at the 20 hr rate

Traction   .   . 1·280 to 1·120–1·150 (130 to 160 points) at the 5-hr rate

It will be seen that the range of specific gravity can vary from 30 points (0·030), for the large stationary cells with heavy plates, to 170 points (0·170) or more for portable cells.

CHANGE IN DENSITY DURING LIFE

Normally the specific gravity of the electrolyte with the battery fully charged does not change appreciably, and there should be no necessity to change the electrolyte or add acid during the useful life of the battery. Towards the end of life, however, there may be a reduction in specific gravity due to partial sulphation of the plates and loss of acid in the active material which falls into the sediment space. *Acid should never be added to a cell during life except to compensate for spilling.* Under normal operating conditions water alone is lost, owing to electrolysis when gassing on charge, and to a slight extent to evaporation. Pure water should be added to restore the electrolyte level before a charge, so that it will mix with the acid when the cell gases during charge.

Age of the Battery

Most batteries are designed to give at least their nominal capacity in the first few discharges. Batteries in the early part of life, particularly those which are *cycled* (given regular cycles of discharge and charge), increase in capacity, and the increase may be as much as 10–20 per cent more than nominal. This higher capacity is maintained for a considerable period, followed by a gradual decline. A battery is usually considered worthless when its capacity has fallen to about 80 per cent of nominal.

Loss of capacity during the life of portable batteries largely results from deterioration of the positive plate by shedding or softening of the active material and corrosion of the grid. The negative plates lose some capacity by hardening of the paste, usually caused by the gradual formation of insoluble lead sulphate.

Voltage

As already stated the voltage of a fully charged storage cell when no current is passing is known as its open-circuit voltage, or electromotive force (e.m.f.).

OPEN-CIRCUIT VOLTAGE AND SPECIFIC GRAVITY

The open-circuit voltage varies with the specific gravity of the electrolyte, and remains substantially constant indefinitely, after a stabilizing period of approximately 12 hr following a charge. This

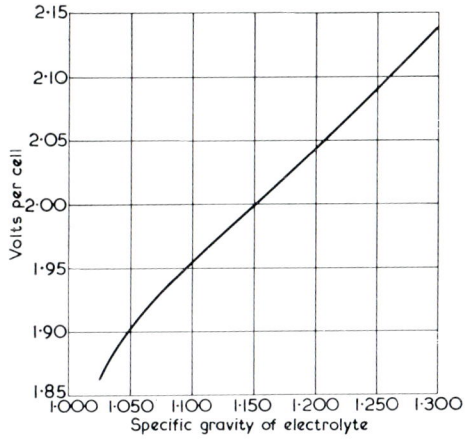

FIG. 3.3. VARIATION IN OPEN-CIRCUIT VOLTAGE OF LEAD-ACID BATTERY WITH SPECIFIC GRAVITY OF ELECTROLYTE

stabilizing period allows the gas or "froth" voltage, evident immediately after a charge, to settle down to the true open-circuit value.

The effect of the specific gravity of the electrolyte on open-circuit voltage is shown in Fig. 3.3, where it will be seen that, over the range of specific gravity from 1·100 to 1·290, the voltage increases uniformly with increase in specific gravity.

Stationary-type cells working in 1·210 sp. gr. electrolyte have an open-circuit voltage of 2·05 V; and automotive cells working in 1·280 sp. gr. electrolyte have an open-circuit voltage of 2·12 V. These figures are typical, so that for most purposes it may be assumed that

$$\text{Open-circuit voltage} = \text{sp. gr.} + 0.84$$

The nominal terminal voltage of a lead-acid cell is taken as 2 V, and of a battery, as the number of cells multiplied by 2.

VOLTAGE ON CHARGE AND DISCHARGE

The voltage of any cell is higher than nominal when it is charging, and lower when discharging. The voltage during charge or discharge varies considerably, being affected by the following factors—

(*a*) Current flowing.
(*b*) State of discharge, or charge.
(*c*) Temperature of cell.
(*d*) Age of cell.

Typical charge and discharge curves are shown in Fig. 3.4 for lead-acid traction-type batteries. These batteries, because they provide electrical power for battery-driven vehicles and industrial trucks, are discharged almost completely each working day or shift. They are recharged, as shown in the curves, in about either 12 or 8 hr, depending on the time available.

Most other types of battery, because they are used as sources of standby or emergency power, do not normally discharge to any great extent and are kept substantially charged most of the time, by working in parallel with a generator or rectifier charger.

In systems where the battery is connected across the charging supply for long periods, the voltage is usually arranged to be below the gassing voltage of approximately 2·30 to 2·40 V per cell, and the charging current passing into the fully charged battery is quite small. Periodically, the battery would be given a gassing charge, or a full recharge whenever more than about 10 per cent of its capacity had been used during an emergency discharge. During such a charge the battery voltage characteristic would be very similar to that shown in Fig. 4.1, on page 51.

VOLTAGE AND CHARGE RATE

For a normal charge the cell or battery is recharged at the rate recommended by the maker, which can be maintained throughout the charge.

Because a discharged battery will safely accept high rates of charge it is possible to commence charge at rates 3 to 5 times that specified for normal charging. This initial high rate of charge must be reduced when the cell reaches gassing point, at approximately 2·35 V, to protect the battery against excessive gassing or rise in temperature.

From the time the cell reaches gassing point to the completion of charge, the charge rate must not be greater than that recommended as the correct "finishing rate." With this method of charging, a fully discharged battery can be recharged in about 6 hr. In service there is normally at least 8 hr available for charging, and as shown in

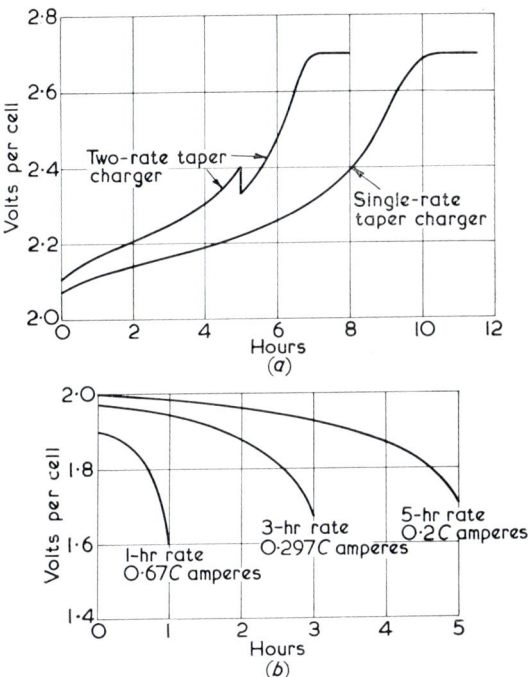

FIG. 3.4. TYPICAL VOLTAGE/TIME CHARACTERISTICS FOR TRACTION CELLS

(a) Charge curves
(b) Discharge curves

C = Ampere-hour capacity at 5-hr rate

Fig. 3.4, a full recharge can be given in this time by commencing the charge at three times the recommended finishing rate.

The effect on the cell voltage of the different charge rates is also shown in the curves. A fully discharged cell will always commence charge at a voltage of at least 2·0 V, and at most 2·20 V, depending on the charge current.

The maximum voltage on charge, obtained with the battery fully charged, varies with the type and age of the battery, and temperature.

Approximate voltages are given below—

Type of battery	Voltage when fully charged (per cell)	Charging current	Datum temperature	
	V	A	deg C	deg F
Traction . . .	2·70	0·07 × C_5	27	80
Automotive and other portable types . .	2·65	0·07 × C_{20}	21	70
Stationary . . .	2·75	0·07 × C_{10}	15·6	60

C_5, C_{20}, C_{10} = Ampere-hour capacity at 5-hr, 20-hr and 10-hr rates

These top-of-charge voltages would be obtained with a finishing rate in amperes equal to 7 per cent of the nominal capacity.

A cell does not always attain the same top-of-charge voltage even at the same charging current as its voltage is affected by temperature, as shown below—

Top-of-charge Voltages at the Finishing Rate

Temperature		Type of battery		
		Traction	Automotive and portable	Stationary
deg C	deg F	V	V	V
15·6	60	2·76	2·68	2·75
21	70	2·73	2·65	2·72
27	80	2·70	2·62	2·69
32	90	2·67	2·59	2·66

The variation in fully charged voltage with temperature for lead-acid cells is approximately 0·03 V for every 5·6° C (10° F), and is deducted for temperatures above, and added for temperatures below, the datum temperature. The fully charged voltage varies also with the charge rate, being higher or lower than the values stated for corresponding changes in the finishing rate.

CHANGES IN CELL VOLTAGE ON DISCHARGE AND CHARGE

As shown in Fig. 3.4, the cell on discharge at the 5-hr rate falls almost immediately from the open-circuit voltage of 2·12 to 2·0 V, owing to the internal resistance of the cell. As the discharge continues, the voltage falls gradually until the amount of lead sulphate formed in the plates and the dilution of the electrolyte result in a fairly steep fall. The cell is considered to be discharged when its voltage has fallen to 1·70 V, at which point the specific gravity of the electrolyte would be approximately 1·140. With the circuit broken and the cell on open-circuit, the cell voltage would recover immediately to almost 2 V. The precise recovery voltage is related to the specific gravity of the electrolyte; Fig. 3.3 shows that at 1·140 sp. gr. the cell voltage would be 1·99 V.

When placed on charge there is an immediate rise in voltage from 1·99 to 2·10 V, mainly due to the sudden increase in the density of the electrolyte in the pores of active material. The subsequent rise of voltage is governed by the rate at which acid is produced in the plates and the rate of diffusion into the free electrolyte of the cell.

The diffusion of acid depends on the viscosity of the electrolyte. This is low at high temperatures and high at low temperatures. At high temperatures diffusion takes place readily and the cell voltage on charge increases slowly. At low temperatures, because of restricted diffusion, the cell voltage is higher and increases quickly with the increasing concentration of acid in the plates.

When the cell voltage on charge reaches about 2·4 V, there is a fairly sharp rise in voltage. At this stage there is almost complete conversion of lead sulphate and the supply of lead ions is reduced. Most of the charge is now being used in dissociating the water of the sulphuric acid solution into hydrogen and oxygen, and the cell begins to gas freely. When this occurs the cell voltage rises less steeply and finally shows no further increase; it may even show a slight decrease if the charge is prolonged so that the temperature of the cell is increased.

Efficiency

The *efficiency* of a battery is the ratio of the output to the input required to restore it to the initial fully charged condition, under specified conditions of discharge and charge rates, and temperature.

As both output and input can be measured in either ampere-hours or watt-hours, efficiency can be expressed as either

$$\text{Ampere-hour efficiency} = \frac{\text{Ampere-hour output}}{\text{Ampere-hour input}}$$

or

$$\text{Watt-hour efficiency} = \frac{\text{Watt-hour output}}{\text{Watt-hour input}}$$

The ampere-hour efficiency of a lead-acid battery is generally about 90 per cent, although for odd discharges where the ampere-hour input is controlled so that very little gassing occurs, the efficiency can be as high as 95 per cent.

The watt-hour efficiency is obtained by multiplying the ampere-hour efficiency by the voltage efficiency—

$$\text{Voltage efficiency} = \frac{\text{Average voltage on discharge}}{\text{Average voltage on charge}}$$

We have seen that the voltage on charge or discharge varies with the current, but typical figures for average discharge voltage of 1·95 V, and average charge voltage of 2·35 V, give

$$\text{Watt-hour efficiency} = \frac{90 \times 1·95}{2·35} = 75 \text{ per cent}$$

Obviously, high rates of discharge or charge will reduce the voltage efficiency and bring about a corresponding reduction in the watt-hour efficiency.

CHAPTER 4

CHARGING METHODS AND EQUIPMENT

CORRECT charging procedure is most important to ensure satisfactory battery life and performance. Modern charging equipment, most of which uses a.c. supply, provides safe and flexible control of battery charging either automatically or with the minimum of manual attention.

Charging Principles

A storage battery cannot supply electrical energy unless the plates have previously been activated by passing direct current through them.

In the manufacture of the lead-acid battery, the raw pasted plates are converted to the active state of lead dioxide (positive plate) and spongy lead (negative plate) in the presence of dilute sulphuric acid during a process known as the *formation charge*.

Following formation the plates are washed and dried and assembled into cells or batteries. After adding dilute sulphuric acid, and before putting into service, it is essential to give the battery an *initial charge*.

INITIAL CHARGE

The initial charge consists in passing a current into the battery for a number of hours as recommended by the battery maker. It is very important that the instructions regarding initial charge be closely followed, as this charge serves to complete the electrochemical conversion of any lead sulphate remaining in the plates. The initial charge ensures that the battery starts its life in the best possible condition so that it is capable of giving rated capacity and satisfactory performance from the moment it goes into service.

CHARGING IN SERVICE

A battery, once it has supplied electrical energy, can be restored to the charged condition only by passing direct current into it—in the correct direction.

The addition of sulphuric acid or patent dopes to a discharged battery cannot replace the electrical energy supplied by the battery.

Correct charging is as important to batteries as correct feeding is to human beings—in either case a deficiency or surfeit will eventually produce ill effects.

Correct charging entails control of the current and time of charge to suit the size of battery and operating conditions so as to ensure the right balance of charge to discharge. This balance is obtained when every cell of the battery is maintained in a healthy condition with the minimum of charge over and above that required to replace previous output.

Obviously, charging methods will be as varied as the duties performed by batteries. A battery used for traction work, where it is substantially discharged almost every day of its life, will be charged very differently from a car battery or a power-station battery, where the load demand is intermittent and of relatively short duration.

There are three main methods of charging batteries in service—

1. *Manual or Semi-automatic Control*

This method applies to all batteries which are disconnected, when discharged, from the equipment they work, so that they can be recharged from a separate charging source. Once the charge has been started by the operator, the charge proceeds without attention and is usually automatically terminated when the battery is charged. The majority of batteries charged by this method are used on battery-electric trucks and vehicles, and operate generally on a daily discharge/charge cycle.

2. *System Control*

Here the batteries are permanently connected to the electrical system, in parallel with a generator. Applications which use this method include motor-cars and vehicles, Diesel-electric locomotives and Diesel units. Once the battery has started the engine, the generator supplies electrical power for the load and for charging the battery. The battery supplies all the load below a certain speed, or when the engine is stopped.

3. *Float and Float/Trickle Control*

This method is adopted for stationary batteries used as sources of standby power in power stations, telephone exchanges and some emergency lighting installations. The batteries are permanently connected to the load in parallel with a charger. In some cases the number of cells is selected so that the charger voltage is just adequate to pass a small charge current sufficient to keep the battery fully charged.

CHARGING TIME AND CURRENT

Where time is an important factor in recharging, as in many industries using battery-electric trucks or vehicles, the shorter is the time available and the greater the average charge current required to replace the energy taken out of the battery.

A battery of 100 Ah which is fully discharged will require between 110 and 115 Ah to restore it to the charged condition. For an 8-hr charge period the average current would be 14 A, and for a 14-hr period, 8 A.

The charge current must comply with the recommendations of the battery manufacturer, particularly towards the end of the charge when the battery commences to gas. This means that an increased charging rate for the shorter charging time of 8 hr or less must be used during the initial stages of the charge.

A discharged battery can safely accept a high rate of charge until it is about 60–80 per cent charged when gassing commences at about 2·4 V per cell. If this initial high rate of charge were maintained beyond this point, only a portion of the charge current would be utilized in charging the battery, and the remainder would be wasted in heating the battery and decomposing the water of the electrolyte into oxygen and hydrogen.

STARTING AND FINISHING RATES

It is essential to reduce the initial rate of charge, or *starting rate*, to the lower recommended value of charging current known as the *finishing rate*. The normal finishing rate for all lead-acid batteries is 6–7 per cent of the nominal ampere-hour capacity. The recommended finishing rate for a 100-Ah battery would be 7 A. Whilst this value should be regarded as the maximum, it is quite permissible and just as efficient to recharge lead-acid batteries at much lower rates down to about 3 per cent if the time available permits.

Charging currents in excess of the recommended finishing rate produce inefficient charging, overheating of the battery, and in time premature shedding of the active material from the plates owing to excessive gassing.

STEP CHARGING

If it is necessary to charge a battery in a short time, it is possible to start the charge at a current 5 or 6 times the finishing rate, i.e. at a current in amperes which is 30 to 40 per cent of the battery capacity in ampere-hours. The current is reduced in stages, or steps, whenever the voltage reaches 2·4 V per cell. The first stage of the charge at 30–40 per cent current would be reduced to half this rate for the second stage. The third and final stage would be carried out at the correct finishing rate until the battery was fully charged.

In the first stage the energy replaced by the time the voltage reaches 2·4 V per cell will be less the higher the charge current. This is shown in the following table for batteries which have been discharged by about 80–85 per cent of their nominal capacity. In practice, however, the starting current need seldom be higher than about 20 per cent of the battery capacity, as the time available for charging is usually 8 hr or longer.

Charging current ratio Amperes / Ah capacity	Time to reach 2·4 V per cell	State of charge at 2·4 V per cell
per cent	hr	per cent
10	8	85
20	3½	75
30	2	65
40	1⅓	57

The charging procedure described above is applicable to batteries which are regularly discharged and charged. In many applications the batteries are not normally deeply discharged; these include batteries used for engine starting in private cars, Diesel locomotives, commercial vehicles, etc. In all these applications the charging current is provided by a generator driven from the main engine, and is automatically controlled by means of a voltage and/or current regulator to suit the condition of the battery. This ensures that a discharged battery receives a fairly high charging current, and a fully charged battery a small charging current, and the control is such that the high starting current tapers down to a safe low value as the battery becomes charged. This method of charging and the equipment used are described in detail in Chapter 6.

Large stationary-type batteries used in power stations as sources of standby power are maintained in a charged condition by being permanently connected across a charging supply. The voltage of the charging supply is only slightly greater than the open-circuit voltage of the battery, so that the battery is merely receiving sufficient trickle charge to compensate for open-circuit losses or any other light loads. These systems are described in detail in Chapter 7.

Charger and Battery Characteristics

When a battery is being charged the current is unidirectional but opposite to that which flows when the battery is being discharged. Most modern battery chargers are operated from a.c. mains, so it is necessary to convert or *rectify* the alternating voltage to direct voltage by some suitable apparatus. The a.c. mains supply is connected to a transformer designed to step the voltage up or down to a value approximating that of the battery to be charged. This alternating voltage is then converted to a suitable direct voltage by means

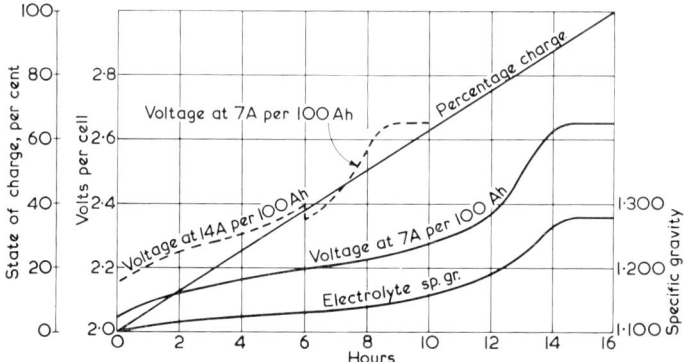

FIG. 4.1. TYPICAL CHARGE CHARACTERISTICS OF LEAD-ACID BATTERIES

of selenium, germanium or silicon rectifiers. In the largest chargers, mercury-arc or silicon rectifiers are used.

In order to pass direct current into a battery during charge, the direct voltage applied to the battery must always exceed the battery back-e.m.f. At the commencement of charge the applied voltage must be between 2·1 and 2·2 V per cell. Thus a nominal 12-V battery will require an applied voltage of about 13·2 V for commencing charge.

As the charge proceeds the battery voltage will increase slowly at first, then fairly rapidly as gassing commences, and will rise to about 2·6–2·7 V per cell, or 16 V for the 12-V battery, towards the end of charge, and the charger must be capable of supplying this output. Typical charge curves are shown in Fig. 4.1, and the following points of interest are worth noting—

(*a*) The difference in voltage and time of charge with the two-rate charging; that is, charging time reduced to about 60 per cent of that for a constant current of 7 per cent.

(*b*) Change in specific gravity of the electrolyte during charge. The specific gravity reading cannot be correlated with state of charge. For example, when the battery is 50 per cent charged (8 hr) the specific gravity is 1·140, whilst in the next 8 hr it increases rapidly to 1·280. This means that the rise is 40 points in the first 8 hr and 140 points in the second 8 hr of charge. There is no uniform rate of increase in specific gravity on charge: the specific gravity reading is significant only in the last hour or so. This is in contrast to the specific gravity of the electrolyte on discharge, where the fall is proportional to the ampere-hours taken out of the battery. With a range of 180 points (1·280–1·100 sp. gr.) for full nominal capacity, the battery would be 50 per cent discharged for a 90-point fall (1·190 sp. gr.), 25 per cent discharged for a 45-point fall, 75 per cent discharged for a 135-point fall, etc.

Many chargers are designed to give a virtually constant direct-voltage supply equivalent to about 2·8 V per cell. A ballast choke and a resistor are usually placed in the transformer or output

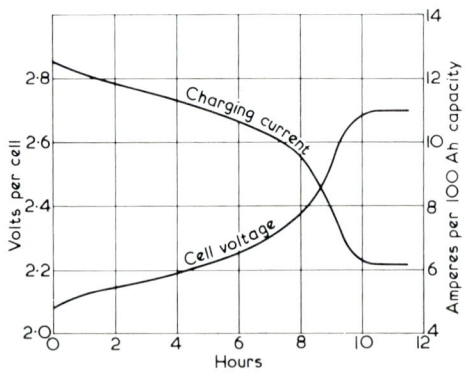

FIG. 4.2. MODIFIED CONSTANT-VOLTAGE CHARGING
Typical charging characteristics showing tapering current with rise in battery voltage

circuit, and these restrict the charging current to a safe value when the opposing battery voltage is low, that is, when the battery is placed on charge in a discharged condition. During the course of the charge the current automatically decreases or tapers as the battery voltage increases (Fig. 4.2).

Charging Methods

Charging methods can be classified under the following headings—

    (*a*) Constant-current charging
    (*b*) Constant-voltage charging

(*c*) Constant-current/constant-voltage charging, referred to as *current-voltage control charging*

(*d*) Modified constant-voltage charging

(*e*) Fast charging

(*f*) Boost charging

(*g*) Trickle charging

(*h*) Float charging

(*i*) Novel systems such as pulse charging, and gas-control charging

CONSTANT-CURRENT CHARGING

The true constant-current method of charging is not necessary for the efficient charging of batteries and is not now widely used. The old notion of keeping the charge current practically constant throughout the course of the charge was useful for quick and accurate determination of the number of ampere-hours put into a battery. It generally entailed frequent manual adjustment of the control resistance in the charge circuit to compensate for rise in voltage of the battery as it became charged. This method is used mostly for the initial charging of batteries, and on the assumption that the current is maintained reasonably constant, the battery manufacturer merely specifies the time required in hours. There would be no objection to specifying the first charge required and the safe current to be used, once gassing commenced. This means that the charging current could commence at up to about twice the specified value, and be reduced to the specified value once the battery voltage had reached the equivalent of 2·4 V per cell. A lead-acid battery can be charged just as efficiently by a taper or falling current as by a controlled constant value of current.

One method of maintaining the charging current more or less constant was to charge from a d.c. supply with a voltage much higher than that of the battery. The charging current was maintained almost constant by using a fixed value of ballast resistance, as demonstrated in the following example.

Nominal voltage and capacity of battery, 12 V, 100 Ah
Supply voltage (V_S), 110 V d.c.
Average charging current (I) to be 7 A

During charge the average battery voltage (V_B) would be 14·4 V (6 × 2·4 V per cell), so that the required series resistance (R) would be given by

$$R = \frac{V_S - V_B}{I} = \frac{110 - 14\cdot4}{7} = 13\cdot6 \ \Omega$$

At the commencement of charge the battery voltage would be 12·9 V (6 × 2·15 V per cell), and at the end of charge, 15·9 V (6 × 2·65 V per cell). Thus

$$\text{Initial current} = \frac{110 - 12·9}{13·6} = 7·14 \text{ A}$$

$$\text{Final current} = \frac{110 - 15·9}{13·6} = 6·92 \text{ A}$$

This method is not to be recommended as it is very inefficient owing to the power (P_R) dissipated in the ballast resistor—

$$P_R = (V_S - V_B)I = (110 - 14·4) \times 7 = 669·2 \text{ W}$$

The power (P) drawn from the mains is given by

$$P = V_S I = 110 \times 7 = 770 \text{ W}$$

so that

$$\text{Efficiency of charge} = \frac{P - P_R}{P} = \frac{770 - 669·2}{770} \times 100 = 13 \text{ per cent}$$

The efficiency of charge using 110-V d.c. mains can be made quite high by increasing the number of batteries.

For example, assuming that six 12-V batteries were connected in series for charging at 7 A from 110-V d.c. mains, the average charging voltage would be

$$V_B = 6 \times 14·4 = 86·4 \text{ V}$$

and

$$V_S - V_B = 110 - 86·4 = 23·6 \text{ V}$$

So that the series resistance to give 7 A is

$$R = \frac{V_S - V_B}{I} = \frac{23·6}{7} = 3·37 \text{ } \Omega$$

If the resistance is fixed throughout the charge,

$$\text{Initial current} = \frac{110 - (6 \times 12·9)}{3·37} = 9·7 \text{ A}$$

and

$$\text{Final current} = \frac{110 - (6 \times 15·9)}{3·37} = 4·3 \text{ A}$$

Although the current starts above and finishes below the recommended rate of 7 A for the size of battery chosen, the method of charging would be quite satisfactory and would give an efficiency of charge of about 80 per cent.

If desired, a variable resistance might be used which could be controlled so that the current was maintained constant—

Maximum resistance at start of charge

$$= \frac{110 - (6 \times 12 \cdot 9)}{7} = 4 \cdot 66 \ \Omega$$

Minimum resistance at end of charge

$$= \frac{110 - (6 \times 15 \cdot 9)}{7} = 2 \cdot 08 \ \Omega$$

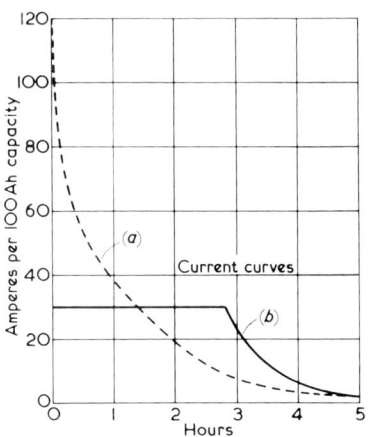

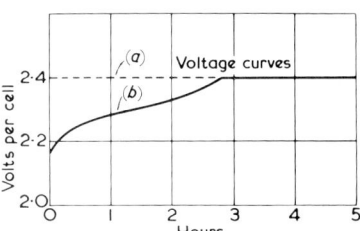

CONSTANT-VOLTAGE CHARGING

In the constant-voltage method of charging, a voltage equivalent to 2·3 to 2·4 V per cell is applied without any ballast resistance in the circuit. Fig. 4.3 shows that a large current passes into the battery initially. As the battery back-e.m.f. increases, the charging current falls fairly quickly, and near the completion of charge, reaches a value less than the finishing rate. This method is quite safe, but true constant-voltage charging is rarely used for batteries which are deeply discharged in service, as the system requires a charger with

FIG. 4.3. RECHARGE CHARACTERISTICS OF LEAD-ACID BATTERIES

(a) From constant-voltage supply equal to 2·4 V per cell
(b) From constant-current supply up to 2·4 V per cell, and thereafter constant voltage

a very high output which is used for a short time only, at the commencement of charge. The method is used mainly for systems where the batteries are not discharged more than about 25 per cent of their capacity. The supply voltage is adjustable between 2·2 and 2·3 V per cell.

It is usual to give periodical equalizing charges to batteries charged by this method, as the system voltage is normally set too low to provide sufficient gassing of the plates to ensure mixing of the electrolyte during normal running. Diesel train units and Diesel-electric locomotives are typical applications using this system.

CURRENT-VOLTAGE CONTROL CHARGING

Constant-current constant-voltage, or current-voltage control charging is widely used for all types of battery applications. Fig. 4.3 shows that initially the current is held at the more practical value of about 20–30 per cent of the battery capacity. When the battery voltage rises to the gassing value of about 2·35 V per cell, the voltage is kept constant at this figure and the current from this point falls quickly as with true constant-voltage charging. This system is widely used for commercial vehicles, and the more expensive private cars whose electrical loads are fairly high. It is also used for parallel charging of traction batteries and is described in greater detail in Chapter 5.

MODIFIED CONSTANT-VOLTAGE CHARGING

A modified form of constant-voltage charging is widely used for traction batteries, which are subject to a regular routine of discharge and charge. A virtually constant applied voltage in the region of about 2·8 V per cell is used, together with a suitable ballast resistor in the charging circuit.

The taper characteristic of the charge current during the recharge of a battery (Fig. 4.2) is typical of most battery chargers which are used with a.c. mains. Further details of these chargers are given in Chapter 5.

FAST CHARGING

The recharging of a battery in a short time, usually less than one hour, is generally restricted to the automotive type of battery. The charge rate is about 80 A for a 6-V, and 40 A for a 12-V battery. Because these high rates of charge are maintained for most of the charging time, the batteries heat up considerably, and the rise in temperature on charge becomes dangerously high if a battery is in an unhealthy condition.

The equipment used for fast charging therefore includes suitable instruments for assessing the condition of a battery before it can be considered suitable for such a charge. Simple tests are carried out on the battery, and if it is proved to be in a satisfactory condition, even though discharged, the charge is switched on and continued for a preset time or until the temperature of the battery reaches 40°– 45° C. At this temperature a thermostat inserted in the centre cell of the battery operates a switch relay which cuts off the charge current.

Fast charging is not intended for regular charging of a battery as it is possible to restore only about 85 per cent of the battery capacity by the time the battery temperature reaches the maximum permissible value. This type of charge also causes wear on the plates due to the excessive gassing which takes place towards the end of the charging period.

Fast charging of automotive batteries is described in greater detail in Chapter 6.

BOOST CHARGING

At certain installations where battery-operated industrial trucks are used it is customary to give the battery a *boost charge* during the lunch break. The normal battery charger is used, so that about 10 per cent of the battery capacity can be replaced in an hour. Inputs up to a maximum of 20–30 per cent of the battery capacity can be obtained in an hour if a current-voltage controlled charger, designed to charge several batteries in parallel, is used for a single battery.

TRICKLE CHARGING

The term *trickle charge* is usually applied to any low-rate charge in amperes equal to 0·05–0·1 per cent of the battery capacity; i.e. 0·5–1 mA per Ah capacity. This trickle-charge current is sufficient to balance the internal losses of the battery, and therefore keeps the battery in a fully charged condition.

Trickle charging has been used extensively for schemes which use Planté-type batteries, and battery life in excess of 25 years has been obtained with this system. It is also used for maintaining other types of battery in a charged condition during fairly lengthy storage periods.

It is advisable to limit trickle charging to a period of 6 months for automotive-type batteries, since the relatively thin grids of the positive plates are liable to deteriorate with continuous low-current charging. Traction-type batteries of either the tubular or heavy flat-plate construction may be maintained for an indefinite period by trickle charge if required.

The trickle-charge current will vary somewhat with the type of plate used or the age of the battery, and the voltage on trickle charge is frequently used as the best guide to correct charging. This voltage should be between 2·2 and 2·3 V per cell, measured not less than 24 hours after placing a fully charged battery on trickle charge.

FLOAT CHARGING

Batteries which are used as sources of standby power and which must be immediately operative in case of emergency must, of necessity, be permanently connected across the system or load. The battery is *floated* across the system in parallel with a rectifier charger, which is voltage controlled to comply with the permissible limits of the system.

Under true float conditions the battery is neither discharging nor charging. In order that it be kept healthy and fully charged, a very small charge current, known as a trickle charge, should flow into it continuously to balance the internal losses of the battery. Under these conditions the battery is operating on float/trickle charge, and the number of cells in the battery is selected so that the applied system voltage is equivalent to about 2·15–2·20 V per cell.

Floating battery systems are widely used for power station, emergency lighting, and telephone exchange installations.

A typical 50-V telephone system with an upper limit of 52 V would have either a 25-cell or 24-cell battery: 25 cells at 52 V would be the equivalent of 2·08 V per cell; and 24 cells at 52 V would be the equivalent of 2·17 V per cell. As the open-circuit voltage of a cell would be 2·06 V, the 25-cell battery would merely float across the system, whilst the 24-cell battery would receive a trickle charge. As the line voltage frequently falls below these values, the battery will slowly discharge, and consequently it must be given periodical charging to restore it to the fully charged condition.

PULSE CHARGING

A method of charging where the charge current is interrupted by the operation of a voltage-sensing relay or a timing device can be used for automatic control charging of batteries which are regularly cycled, and also for maintaining charged batteries in good condition.

The recharge of a discharged battery proceeds normally until the battery voltage attains the equivalent of about 2·5 V per cell, when the voltage relay operates and terminates the charge. With the charge cut off, the battery voltage falls instantaneously to about 2·3 V per cell and thereafter at a slow rate. After half an hour or so, when the voltage has fallen to about 2·2 V per cell, the charge is restarted automatically. The battery voltage builds up in a few minutes to the equivalent of 2·5 V per cell, when the whole cycle of operations is repeated.

The time ratio of charge "off" to charge "on" is about 6 to 1, so that under normal conditions the battery can be left connected to the charger until required.

The dangers of charge termination by voltage for applications where the battery is regularly cycled are the effects of temperature and age on battery voltage on charge. Increasing temperature and age produce a reduction in the top-of-charge voltage, and it is possible for batteries to fail to reach the voltage to be sensed by the relay so that pulsing does not occur and there is a grave danger of the battery being over-charged.

In the case of charged batteries being maintained in good condition it is usual for the pulse charger to be controlled by a time switch so that the "off" to "on" time ratio is about 60 to 1, that is, the charge current is timed to be on for 1 minute in every hour or 1 second in every minute. This method has little or no advantage over the method of maintaining batteries by trickle charging.

GAS-CONTROLLED CHARGING

A new and interesting development for the control of charge of traction batteries is *gas-controlled charging*. This involves the connexion of a heat-sensing device in the control circuit of the charger. The heat is produced in a special vent plug containing a catalyst, and one or two of these plugs are fitted to the cells of a battery. The heat is generated in the vent plug by the recombination of the hydrogen and oxygen gases given off by the cell once the voltage reaches approximately 2·4 V. From this point the charge current is continuously controlled by the heat produced in the vent plug, and charging continues at a safe gassing current. By careful choice of the heat-sensing device it should be possible to charge a battery without producing an excessive amount of gas in the cell, resulting in less wear and tear on the plates.

Alternatively, the heat produced by the catalyst in the special vent cap can be used to operate a thermal switch which interrupts the charge. When no further gas is evolved from the cell the vent cap and thermal switch cool down and charging is restarted, so that a series of pulse charges is produced.

Both systems depend on the complete reliability of the special "catylator" plug, and it is usual to fit more than one of these to a battery. In the event of one failing, the other plug working in parallel takes over. So far, complete reliability of the special plug has not been proved.

Charging Equipment

Few, if any, d.c. mains remain in this country, and almost all battery-charging equipment is supplied by alternating current, which, strange as it may seem, is more adaptable for battery charging than direct current.

The production of direct current by means of an alternating voltage, is known as *rectification*, and the device which brings this about is called a *rectifier*.

PRINCIPLES OF RECTIFICATION

An alternating-current supply is constantly changing in magnitude and reverses its polarity periodically. With a 50 Hz a.c. supply this occurs once every 1/100 sec for in 1 sec there are 50 pulses in each direction. If a battery were connected across such a supply the positive pulses would tend to charge the battery, and the negative pulses to discharge it. If, however, a rectifier is put in one leg of the a.c. supply it will allow current to pass in one direction only. Such a rectifier provides *half-wave rectification*.

FULL-WAVE RECTIFICATION

Practically all battery chargers are designed to give *full-wave rectification*, which is commonly produced by bridge-connected semiconductor rectifiers. A typical bridge circuit is shown in Fig. 4.4.

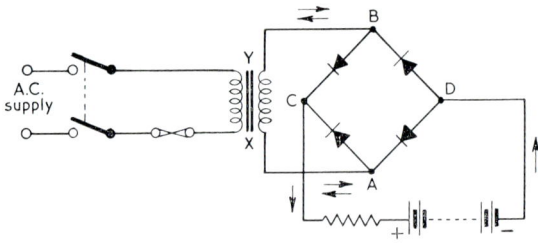

FIG. 4.4. BRIDGE-CONNECTED RECTIFIERS FOR FULL-WAVE RECTIFICATION

The rectifiers allow current to flow in one direction only, as indicated by the vertical arrows. When X is positive the current flows from A to C, through the resistor and the battery, from D to B, back to Y. When Y is positive the current flows from B to C, through the resistor and battery, from D to A, back to X. The resulting voltage waveform, shown in Fig. 4.5, gives 100 pulses per second in the positive direction. If we superimpose the battery voltage it will be appreciated that a charging current will pass into the battery only when the impressed voltage exceeds the battery voltage. The shaded areas indicate when this occurs; the resulting charging-current waveform is also shown.

The average value of this current, as read on a moving-coil ammeter, is the effective current for charging a battery. With a pulsating charging current, a moving-iron type of ammeter does not measure the average current but the heating current, which is always greater than the average current. For this reason it is desirable to use only moving-coil ammeters for rectifier chargers.

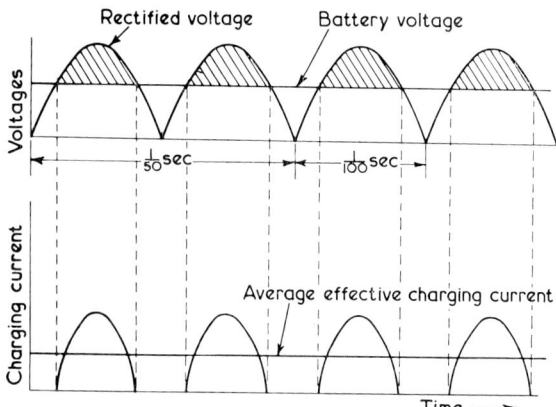

FIG. 4.5. VOLTAGE AND CURRENT WAVEFORMS WITH FULL-WAVE
RECTIFICATION

RECTIFIERS

Many kinds of rectifier have been used for battery charging. The more important ones include rotary convertors, mercury-arc rectifiers, thermionic valves, and semiconductor rectifiers; the latter include metal rectifiers (cuprous oxide and selenium) and junction rectifiers (germanium and silicon).

The rapid development, convenience and efficiency of static rectifiers using selenium, germanium or silicon elements has resulted in these types being used almost exclusively for battery chargers.

The first metal rectifier using a combination of copper and cuprous oxide was used for battery chargers as far back as 1930. By the early 1940s the selenium rectifier had replaced many of the cuprous-oxide types, because of superior output and higher efficiency for less bulk. By the early 1950s considerable progress had been made with germanium rectifiers. These were much smaller than the selenium rectifier for the same current rating and had the advantage of not ageing in service. This ageing effect, where the rectifier output falls after a period in service, had been particularly noticeable with many early types of selenium rectifier used for traction-battery chargers, and by the late 1950s the germanium rectifier was being widely used in the larger battery chargers. By 1958, a rectifier superior in many characteristics to either selenium or germanium had been developed using silicon, and although its use in battery chargers is rather limited at present, it is likely to be used extensively in the future, as silicon is more readily obtainable than germanium, and can operate at a higher temperature.

The future and scope of the various types of rectifier in the field of battery charging seem likely to fall in the following categories—

(*a*) Selenium rectifiers for most small battery chargers of less than 1 kW.

(*b*) Germanium rectifiers for chargers of 1–10 kW.

(*c*) Silicon rectifiers for chargers of over 10 kW, ultimately replacing mercury-arc rectifiers.

RANGE OF BATTERY CHARGERS

Battery chargers are made in a very wide range of voltage and current output to suit any number of cells and capacity. The greater the capacity of the battery to be charged, the higher must be the current rating of the charger, for any given charging time. Most traction-battery chargers are designed to charge a battery in not less than 8 hr safely. The majority are designed to recharge a battery fully in 12–14 hr, as the battery is not usually required for more than 8–10 hr per day. Once the battery has been connected to the charger, charging is entirely automatic, and the current is switched off when the charge is complete. Some of the largest chargers are used by the battery manufacturers for initial charging of plates and batteries before they leave the works. The latest designs use germanium or silicon rectifiers working in parallel to supply a direct current of more than 1,000 A at 110 V.

At the other end of the scale are the small home chargers suitable for charging several cells, or a 6-V or 12-V car battery, at one or two amperes. These car-battery chargers should be required only for the odd occasion, when the battery has been deeply discharged by leaving lights on for several hours with the car parked, or other similar occurrence. With normal running the generator and regulator of the car should maintain the battery in a satisfactory state of charge.

CHARGER OUTPUT CONTROL

In most of the large chargers, the charging current and the termination of the charge when the battery is fully charged are controlled automatically. In the relatively simple home chargers, where the output is limited to one or two amperes, control of either current or time is not so important and the charge must be terminated by switching off the main supply.

The control of charger output is effected by one or other of the following methods—

    (*a*) Resistance or choke control
    (*b*) Transformer tappings
    (*c*) Grid control of mercury-arc rectifiers
    (*d*) Gate control of silicon controlled rectifiers

Most chargers incorporate a ballast choke or resistor in one of the output leads to the battery. Where the charger is designed to charge a definite number of cells a fixed value of resistance is used. This gives a taper or decreasing charge current as the battery voltage increases on charge. Where a charger is designed to charge a variable number of cells the resistor is made adjustable to cater for the different battery voltages which may be charged from the one charger.

In place of, or in addition to the ballast resistor, an iron-cored choke may be connected in series with the primary or secondary winding of the transformer. Owing to its inductive effect, the choke is a more efficient form of circuit ballast than a pure resistor. A charger with choke control will have a higher efficiency than one with resistance control and will be less affected by changes in mains voltage.

TRANSFORMER TAPPINGS

In certain chargers it is possible to select various tappings on the transformer secondary winding by means of a rotary switch. This provides an increased voltage across the rectifier to compensate for rise in battery voltage during charge, without loss of power in the form of heat as obtains with resistance control.

GRID CONTROL OF MERCURY-ARC RECTIFIERS

The charging current can be varied by the voltage applied to the grid of a mercury-arc rectifier. Control of the instant at which the voltage is applied to the grid can delay the firing of the rectifier, thus reducing the effective charging current.

Operation of the control to the limit, so that firing is unduly delayed, produces a low power factor and a peaky output waveform, and may in some applications be detrimental to the battery.

SILICON CONTROLLED RECTIFIERS

This latest development in the control of charging current by means of silicon controlled rectifiers is most interesting. The basic circuit is shown in Fig. 4.6. The bridge-connected rectifier has two arms containing standard silicon rectifiers, and two containing silicon controlled rectifiers.

Either of the *silicon controlled rectifiers* (*SCR*) opposes the flow of current in both directions until a low-power input signal is fed into

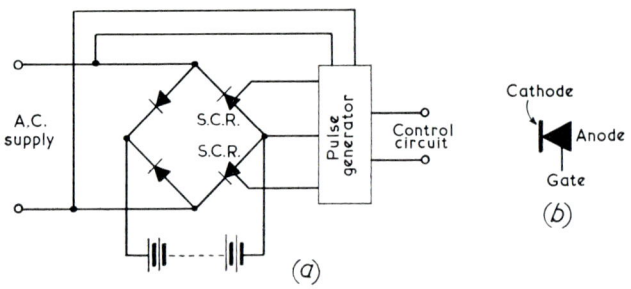

FIG. 4.6. SILICON CONTROLLED RECTIFIER CHARGING
(*a*) Basic circuit
(*b*) Graphical symbol for silicon controlled rectifier (S.C.R.)

the control electrode, or *gate*. It then behaves like a normal rectifier and passes current for, say, the forward half-cycle. During the reverse half-cycle the second silicon controlled rectifier is "fired" and allows current to pass.

As in the grid control of mercury-arc rectifiers, the mean charging current passing into a battery may be varied by controlling the instant in every half-cycle at which the silicon controlled rectifier is "fired."

METHODS OF TERMINATING CHARGE

Small home chargers, because they are relatively cheap and designed to provide only relatively low charging currents, are not fitted with a cut-off device. This means that the charger must be disconnected from the mains supply when the battery has been charged.

Medium-size chargers intended for recharging the largest automotive, or smallest traction, batteries are frequently fitted with a current relay which operates and terminates the charge when the current has fallen to a certain value. This is usually about 60 per cent of the charger output. and corresponds to a battery voltage of 2·5 V per cell. Variations in mains voltage produce corresponding

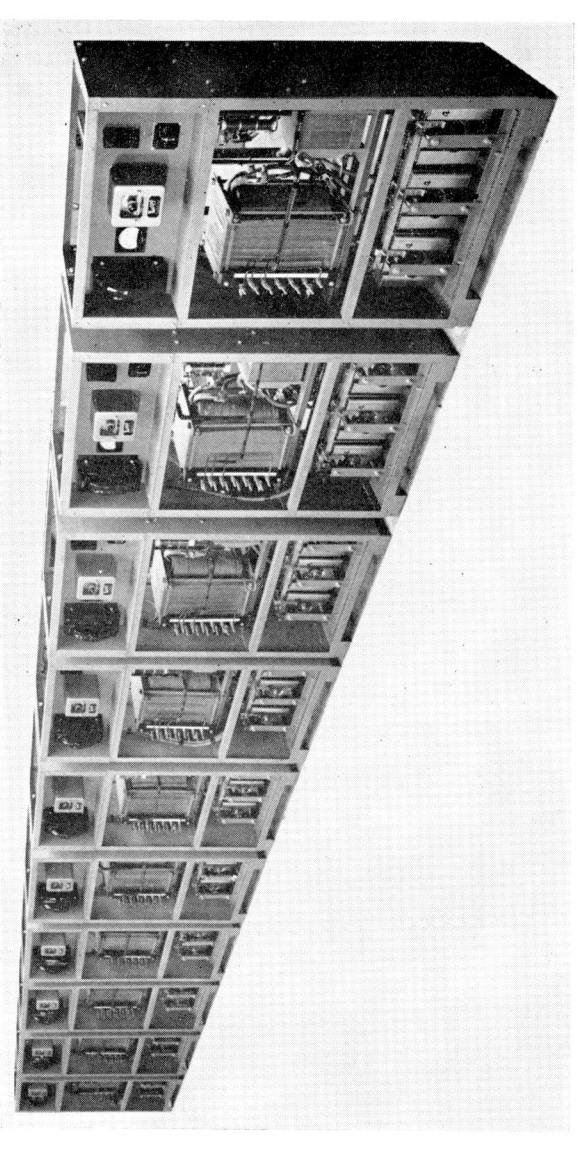

Fig. 4.7. TRACTION-BATTERY CHARGERS

Bank of chargers for charging 18-cell 960-Ah batteries in 8 hr (front panels removed to show: germanium rectifiers (*bottom*); transformer, choke and resistor (*middle*); ammeter, fuses, control switch and voltage/time relay (*top*)

(*Legg (Industries) Ltd.*)

wider variations in charging current which affect the operation of the current relay and make the termination of charge by this method rather indeterminate.

The larger chargers, and particularly those used for traction batteries (Fig. 4.7), use a voltage relay set to operate a timing mechanism when the battery voltage on charge reaches 2·35 V per cell. The timing mechanism can be varied between 0 and 6 hr to suit any particular battery and operation. The principle and details of this device are described in some detail in Chapter 5.

CHAPTER 5

TRACTION BATTERIES

TRANSPORTATION of men and materials in the shortest possible time, and with the minimum of human effort, has been a challenge which has stimulated inventors through the ages. The discovery about the year 1830 of rotary motion by electromagnetism, with batteries as the source of electrical power, provided a new and exciting field of investigation into propulsion which was both silent and clean compared to steam.

The early trials with battery railcars were handicapped by the limitations of the primary battery. Some of these limitations were removed towards the end of the nineteenth century by the rapid development of lead-acid batteries, which provided greater electrical power and longer life than the best of the primary batteries.

In many countries, tramcars driven by lead-acid batteries were introduced—the largest car carrying up to thirty passengers over a range of fifty miles per battery charge. About the same time, battery-propelled taxis and private cars appeared on the roads to rival steam cars. Later the introduction and rapid development of the internal-combustion engine, together with the expansion of the petrol and oil industries, combined to oust battery-electric vehicles for most types of transport.

There remain, however, several fields of application where the battery-electric vehicle or truck is firmly established, and, in fact, being used increasingly year by year.

Battery-Electric Vehicles

About the time when battery-driven trams were becoming obsolete, battery-driven vehicles were introduced in many towns for the delivery of milk, bread, etc., and the collection of refuse. This type of transport is still widely used today, and some of the characteristics of the modern battery-electric road vehicles and industrial trucks which make them attractive compared to vehicles powered by internal-combustion engines, are listed below.

1. Operational costs are lower.

   (*a*) Electrical power is cheaper and at a more stable price than either petrol or oil, which are taxed. Vast improvements in battery life, where the guaranteed life has increased from 2 years to 4 years with an average life expectancy of almost 6 years, have contributed to lower "fuel" costs.

(*b*) The battery-electric vehicle requires much less maintenance, and it has far fewer moving parts than the internal-combustion engine.

(*c*) Economy is greatest on frequent stop-start short-distance service. This is important with the present-day dense traffic in town and urban areas, entailing frequent traffic stops in addition to normal service stops. An electric vehicle consumes no power when stationary, and its economical speed is adequate for the areas served.

(*b*) The life of the electric vehicle is about 12 years, compared to 6 for the motor vehicle.

2. It is simple to drive, and is inherently safe.

3. It is relatively silent and gives off no noxious fumes, so that it is particularly suitable in food stores, warehouses, coal mines, tunnels, and hazardous atmospheres.

4. New designs of traction batteries give up to 35 per cent more capacity, for the same volume and weight, than the older types. Developments in electric motor control by silicon controlled rectifiers in place of resistance control should economize in battery size and increase the range of the vehicle.

BATTERY-ELECTRIC INDUSTRIAL TRUCKS

Battery-electric trucks include all types of industrial truck designed to perform a wide variety of duties. Since their introduction, industrial battery trucks have made a major contribution to industrial efficiency by providing improved materials handling, in speeding up movement of raw and finished materials, and in saving valuable space and manpower.

TYPES OF TRUCK

The main groups of battery-powered industrial trucks used for transporting, lifting and stacking material are

(*a*) Platform
(*b*) Pallet
(*c*) Fork-lift
(*d*) Tractor

Types (*a*)–(*c*) may be either pedestrian- or rider-controlled.
The platform truck is one of the earliest types of works truck. The

platform may be either fixed or elevating, the latter type enabling a preloaded "skid" to be raised from the ground, transported and deposited without waiting time. A development of the platform truck is the pallet truck, on which a metal pallet capable of holding loads up to 2540 kg is used.

The fork-lift truck may be of either the counterbalanced type or the reach type. The fork-lift truck operates by raising preloaded

FIG. 5.1. FORK-LIFT TRUCK, COUNTERBALANCED TYPE
The battery is housed in a steel tray in a compartment below the driving seat
(*Conveyancer Fork Trucks Ltd.*)

hollow box platforms, or pallets, by means of two horizontal steel arms, or forks, which slide into the pallet.

In the counterbalanced type (Fig. 5.1), the forks extend beyond the front wheels, so that, when lifting and carrying a heavy load, the counterbalance weight behind the mast of the truck body and battery must be adequate to provide stability when running and turning. There is a very wide range of fork-lift trucks available which are capable of lifting from 500 to 3500 kg.

The reach truck (Fig. 5.2) is a comparatively new development designed to save space when stacking and storing material. It can operate in narrow working passages, and differs from the counterbalanced type in its more compact size. This is achieved by the

FIG. 5.2. FORK-LIFT TRUCK, REACH TYPE
The battery compartments are on either side of the mast
(*Lansing Bagnall Ltd.*)

movable forks withdrawing the load within the wheel base, and by using tall two-piece batteries mounted on either side of the mast. The popularity and demand for the reach truck are certain to increase as economy of warehouse space becomes increasingly important.

Tractors are designed to pull or push one or several trailers, and the largest are capable of moving loads of several tons.

BATTERY-ELECTRIC ROAD VEHICLES

The type of light delivery van shown in Fig. 5.3 is widely used for the delivery of bread, milk, etc. Such vans are capable of speeds

FIG. 5.3. MILK DELIVERY VEHICLE
The battery is in a wood tray between the wheels
(*Smith's Delivery Vehicles Ltd.*)

up to 32 km/hr and have a loading capacity of 500 to 2000 kg. Heavier and larger vehicles are used for refuse collection. The growth of large flats and compact housing estates has created a demand for battery-operated "prams" for milk delivery. These operate over a relatively small area and are pedestrian controlled at a speed of about 5 km/hr.

BATTERY LOCOMOTIVES

Battery locomotives are designed to run on tracks, and are therefore steel-tyred. They are used in railway, industrial, mining and tunnel applications.

For use in coal mines or other hazardous atmospheres, all the electrical equipment must be flameproof, and covered by a certificate issued by the Ministry of Power. Fig. 5.4 shows a 14,000 kg approved-type battery locomotive designed for underground coal haulage. The 200-V battery is of 100-kWh capacity, and consists of 100 Exide Ironclad Gauntlet cells of 504 Ah. The battery tank is located between the two driving cabs, and is mounted on rollers for easy removal. The National Coal Board operates between 350 and 400 battery locomotives for haulage underground.

FIG. 5.4. APPROVED-TYPE BATTERY LOCOMOTIVE (14 TON) FOR
UNDERGROUND COAL HAULAGE
The battery tank is between the driving cabs
(*Greenwood & Batley Ltd.*)

TRACTION BATTERY TYPES AND DESIGN

Lead-acid batteries are the type most widely used for motive purposes, although, if desired, batteries of the nickel-cadmium and nickel-iron alkaline types can be supplied; these are much more costly but generally have a longer life than lead-acid batteries. The characteristics of alkaline batteries are discussed in Chapter 8.

The type of lead-acid battery used for traction work is characterized by ruggedness of design to withstand service conditions imposed by the hardest of all battery applications. The construction differs from that of other pasted-plate batteries mainly in the positive plates, which may be tubular or of the heavy flat pasted type. Both these types of positive plate are used in conjunction with a flat pasted-type negative plate. Their general construction is shown in Figs. 5.5 and 5.6.

The Exide-Ironclad cell of Gauntlet construction (Fig. 5.5) has a positive plate consisting of a multi-tubular sleeve of Terylene suitably treated and stiffened. The Terylene sleeve is threaded over a conducting antimonial-lead grid, built up of a number of lead spines held in a vertical frame by a top casting of antimonial lead and a

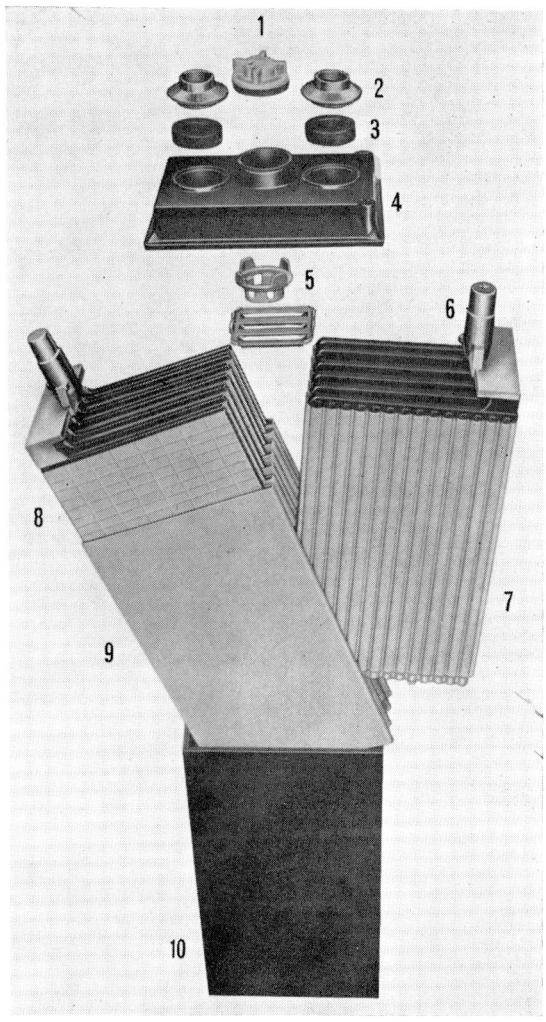

FIG. 5.5. EXPLODED VIEW OF EXIDE-IRONCLAD TRACTION CELL WITH "GAUNTLET" PLATE CONSTRUCTION

1. Vent plug
2. Lead post rings
3. Rubber sealing gaskets
4. Cell lid
5. Separator guard

6. Terminal posts
7. Positive plates
8. Negative plates
9. Porvic sleeves
10. Cell box

bottom bar of moulded polythene. The space between the lead-alloy spines and the Terylene tubes is packed tight with active (positive) material. This type of positive plate construction gives great strength and a high degree of porosity for acid penetration. Terylene is highly resistant to sulphuric acid, and combines strength and elasticity with high resistance to oxidation. Negative plates of the flat pasted type are enclosed by microporous Porvic sleeves in such a way that short-circuits across the plate edges are eliminated.

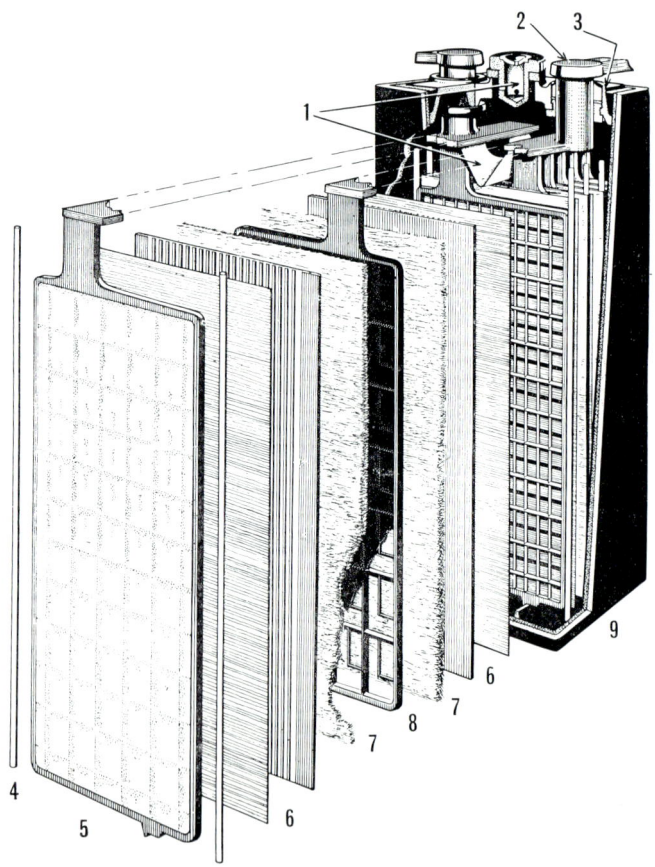

FIG. 5.6. EXPLODED VIEW OF FLAT PLATE TRACTION CELL

1. Spray arrestor and vent plug
2. Intercell connector
3. Sealing gland
4. P.V.C. rod
5. Negative plate

6. Porvic and wood diaphragm separators
7. Glass-wool separators
8. Positive plate
9. Hard-rubber box

The tubular type of traction battery using slitted ebonite tubes has largely been replaced by the type of battery described above.

The Pg type of tubular plate uses a double sleeve of woven glass fibre and perforated polyvinyl chloride (p.v.c.).

The traction type cell shown in Fig. 5.6 uses a flat pasted positive plate, designed to give long life under arduous working conditions. The paste of the positive plate is firmly supported by the pressure of a mat of glass-wool, used together with a ribbed Porvic separator and a treated wood veneer adjacent to the negative plate. P.V.C. rods are fitted down the edge of the negative plate to provide insulation and prevent internal short-circuits.

Battery Life

Many of the traction battery applications outlined in the early part of the chapter usually demand a daily routine of discharge and charge. Although operating conditions vary considerably, it is usual to obtain an average battery life in excess of the four years guaranteed by the battery manufacturer, and under fairly normal conditions, lives between five and six years are common. Under these conditions the battery is not regularly required to provide more than about 85 per cent of its 5-hr capacity. Occasional discharges to the full amount of the battery capacity do no harm, provided that recharging is prompt and adequate.

Persistent overdischarging, that is, discharging beyond the capacity represented by the specific gravity reading recommended by the battery manufacturer, may lead in time to deterioration. When a battery is overdischarged the excess of lead sulphate in the plate produces over-expansion of the active material, so that the pores become closed and it becomes difficult to reconvert all the sulphate on the subsequent recharge completely.

Battery Sizes and Output

To meet the large variety of applications, traction cells vary considerable in physical size, weight and capacity. The smallest cell, 24 cm tall, volume 2·5 dm³, weight 5·5 kg, has a capacity of approximately 50 Ah. One of the largest cells available, 74 cm tall, volume 49 dm³, weight 145 kg, has a capacity of 2,000 Ah. The capacities per unit weight and volume of the traction battery, complete with outer tray, are approximately 24 Wh per kg and 55 Wh per dm³. These values can be improved by using thinner plates, but only at the expense of battery life under the arduous operating conditions of traction work.

Choice of Battery

Factors which determine the choice of battery for traction applications are—

1. Nature of duty.
2. Size and voltage of electric motors chosen for the truck.
3. Space available for the battery.

Motors used in most trucks and vehicles vary from about 2 hp to 12 hp, and industrial fork-lift trucks will have at least two motors, one for propulsion, the other for working the pump for hydraulic lift.

Voltages vary from 12 V (6 cells) to 64 V (32 cells) for industrial trucks, and from 30 V (15 cells) to 72 V (36 cells) for dairy and bakery road vehicles. Truck and vehicle manufacturers will offer users various batteries for the same truck or vehicle, to meet light, medium or heavy duties. The average industrial truck battery, of approximately 10 kWh, would weigh about 500 kg, would occupy 210 dm³, and would normally work an 8-hr shift per charge. Where trucks are required for more than eight hours it is usual to fit oversize batteries or to change batteries at eight-hourly intervals.

Batteries in road vehicles vary considerably in capacity and voltage. Factors which affect the choice of battery are: payload, distance to be covered, number of stop/starts, and nature of the roads. A vehicle which travelled a specified distance with a defined number of service stops on fairly level ground would require a battery of about 30–40 per cent more capacity to do the same duty in hilly districts.

Where intending customers are in doubt regarding choice of vehicle and battery, truck and battery manufacturers are prepared to carry out a practical road survey simulating actual service conditions, so as to assess the correct combination of vehicle and battery for the proposed duty.

Battery Assembly and Mounting

A traction battery consists of a number of individual cells, connected in series, assembled in a crate or outer container of wood or metal.

Intercell connectors of lead-plated copper strip, insulated flexible cable, or solid lead-alloy are burned (welded) or bolted to cell terminal posts. In the interests of maintaining sound electrical connexions when subjected to bumps and vibration in service, the connectors are usually solidly burned to the cell posts.

As battery space on most trucks is restricted, the physical size of each cell is kept to a minimum by using thin-walled cell boxes whilst ensuring a high resistance to shock by the use of box material in new resin-rubber compounds.

For road vehicles it is usual to assemble cells in wood trays.

Most industrial truck batteries are assembled in metal containers in which the cells are packed direct with packings of a suitable material inserted between the cells and the walls of the container. A typical industrial truck battery in a metal tray is shown in Fig. 5.7.

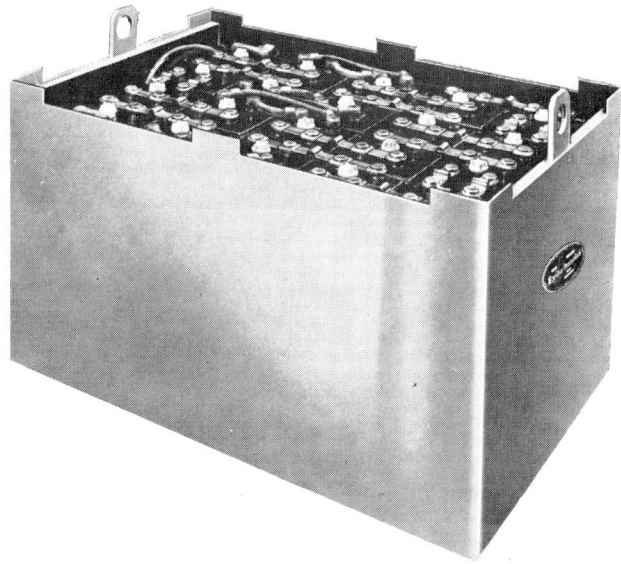

FIG. 5.7. INDUSTRIAL TRUCK BATTERY IN METAL TRAY

In all battery assemblies, the cells are packed tightly into the outer container, which has to be thick and strong enough to withstand rough service conditions and the internal pressure exerted by the cells owing to expansion of the active materials during life. It is often necessary to reinforce containers by means of internal partitions and external tie rods.

To facilitate ease of handling of heavy batteries, all wood trays are provided with lifting irons, and metal trays with lugs, so that chains can be attached for lifting the container on or off the truck by means of a small hoist or fork-lift truck. Almost all battery trays or containers are fitted with removable or hinged lids which serve as a cover and protection for the cells against dirt and moisture, and prevent metal objects being dropped across the cell tops with the danger of short-circuiting of live connectors. The battery lids are removed during charge to assist ventilation, and prevent undue temperature rise within the battery.

Discharge Characteristics

The curves of battery performance shown in Figs. 5.8 and 5.9 are typical of all lead-acid traction batteries, whether of tubular or flat plate design.

A battery reacts to work very similarly to a human being—too much output at a time and it quickly becomes exhausted. This is

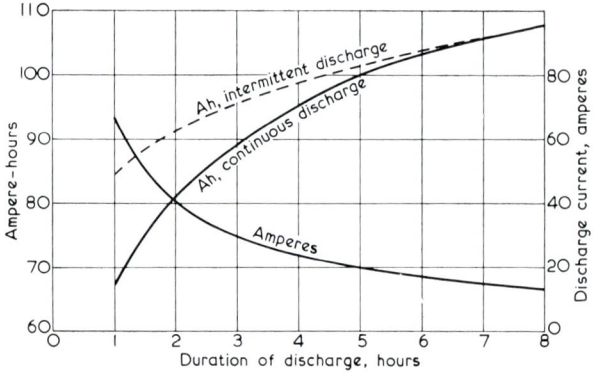

FIG. 5.8. CAPACITIES OF A TRACTION BATTERY ON CONTINUOUS AND INTERMITTENT DISCHARGE AT NORMAL TEMPERATURE (27° C OR 80° F)

Nominal capacity, 100 Ah on continuous discharge at 5-hr rate

demonstrated by the capacity available at the various rates of discharge shown in Fig. 5.8 for a battery of 100 Ah at the 5-hr rate of discharge at 27° C (80° F)—

8-hr rate: 108 Ah (or 13·5 A for 8 hr) available
5-hr rate: 100 Ah (or 20 A for 5 hr) available
3-hr rate: 89 Ah (or 29·7 A for 3 hr) available
1-hr rate: 67 Ah (or 67 A for 1 hr) available

The above figures represent the capacity available when the discharge current is continuous for the time under consideration. In service, where batteries are driving trucks and vehicles on a stop/start/rest schedule, the discharge is intermittent and is usually spread over a 6- to 8-hr period.

The discharge current taken from a battery to supply most of the motors used on electric trucks is as high as the 2-hr rate for the size of battery used. As the discharge is intermittent, practically the full 5-hr capacity is available. The curve in Fig. 5.8 shows that 91 per cent of nominal capacity is available when the 2-hr current is taken intermittently over a period of 6 hr or so. This compares with 81 per cent of nominal capacity available when the 2-hr current is taken continuously.

The gain in useful capacity by discharging intermittently and the reduction in available capacity at low temperatures (Fig. 5.9) is

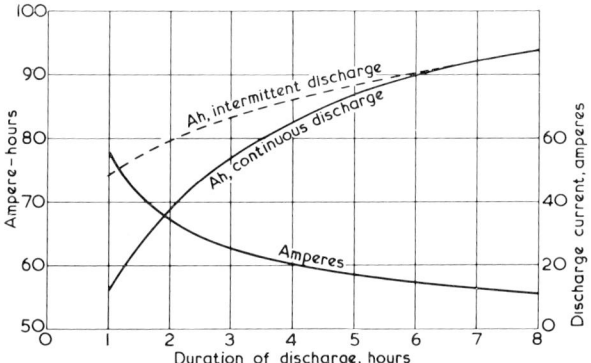

FIG. 5.9. CAPACITIES OF A TRACTION BATTERY ON CONTINUOUS AND INTERMITTENT DISCHARGE AT LOW TEMPERATURE (0° C OR 32° F)

Nominal capacity, 100 Ah on continuous discharge at 5-hr rate and normal temperature

mainly the result of electrolyte diffusion. Intermittent discharging allows time for diffusion to take place, whilst reduction of battery temperature restricts diffusion because of the increased viscosity of the electrolyte.

Comparing the capacity available on continuous discharge at 27° C (80° F) and 0° C (32° F), we have

Rate of continuous discharge	Capacity available	
	27° C	0° C
hr	Ah	Ah
5	100	87
3	89	77
1	67	56

Comparing Figs. 5.8 and 5.9, it will be observed that when discharging *intermittently* the 1-hr capacity (74 Ah) at 0° C (32° F) is higher than that obtained (67 Ah) on *continuous* discharge at 27° C (80° F).

The power in watts available from a battery is the product of volts and amperes, and for an electric motor of given size there are alternative ways of providing the requisite power from a battery.

For example, a 5-hp motor, with allowance for inefficiencies, would take a current of approximately

5,000 W/36 V = 139 A from an 18-cell battery, or

5,000 W/48 V = 104 A from a 24-cell battery

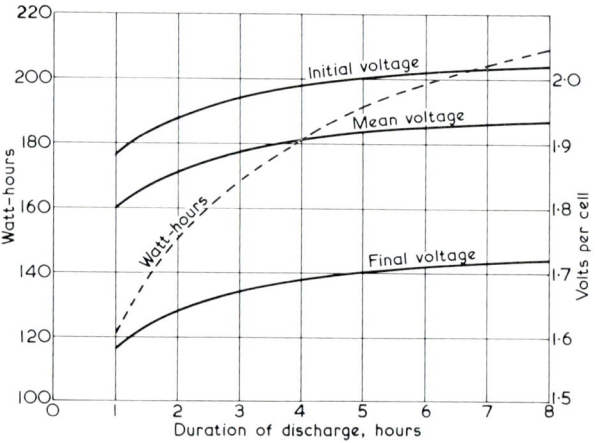

FIG. 5.10. VOLTAGE AND POWER OUTPUT OF TRACTION BATTERY AT NORMAL TEMPERATURE (27° C OR 80° F)

Nominal capacity, 100 Ah at 5-hr rate

If the current was required continuously for 2 hr the ampere-hour capacity required would be

2 × 139 = 278 Ah for the 18-cell battery, or

2 × 104 = 208 Ah for the 24-cell battery

Using the data given in Fig. 5.9, where the 2-hr capacity is 81 per cent of the 5-hr capacity, the size of battery required, related to the nominal 5-hr rating, becomes

$$18 \text{ cells of } 278 \times \frac{100}{81} = 343 \text{ Ah, or}$$

$$24 \text{ cells of } 208 \times \frac{100}{81} = 257 \text{ Ah}$$

Cells nearest in capacity to those calculated above would be selected from battery catalogues, and the final choice as to the height and base area of the cells would be determined by the space available in the truck.

To determine the power (watts) or energy (watt-hours) output of a battery, the average operating voltage of the battery must be known. Fig. 5.10 shows the average discharge voltage per cell at various rates of discharge. This is always less than 2·0 V at traction discharge rates, and decreases with increase in discharge current. The watt-hour output of a cell is therefore the product of the average discharge voltage and the ampere-hours available at the particular rate of discharge, as shown in the following table for a 100 Ah cell—

Rate of discharge	Capacity available	Average voltage per cell	Energy output
hr	Ah	V	Wh
5	100	1·92	192
3	89	1·885	168
1	67	1·80	121

The end, or final, discharge voltage is taken as 0·3 V per cell lower than the initial discharge voltage. This is the point in the discharge characteristic where the cell voltage commences to fall fairly quickly and no further useful power is available. The discharge-voltage/time curves of Fig. 3.4 demonstrate the fairly sharp bend or knee in the voltage curve at this point.

Temperature

Changes in battery temperature produce corresponding changes in performance: a rise in temperature increases the available capacity, and a fall in temperature reduces it. This is a temporary effect only, and capacity is restored with a return to normal temperature.

Lead-acid traction batteries can and do operate quite satisfactorily in cold stores and cold climates, with ambient temperatures as low as $-29°$ C ($-20°$ F), provided that the battery temperature at the start of operation in the cold atmosphere is about $25°$ C ($75°$ F) or higher.

82   *Traction Batteries*

A battery does not readily lose stored heat, and Fig. 5.11 shows the rate of cooling of a truck battery of 18 cells and 440 Ah, housed in a steel tray without any thermal insulation. The battery operated a fork-lift truck over a period of 8-hr in a cold store at an ambient temperature of $-29°$ C ($-20°$ F). The battery was still operating the truck without difficulty at the end of the 8-hr period, and would

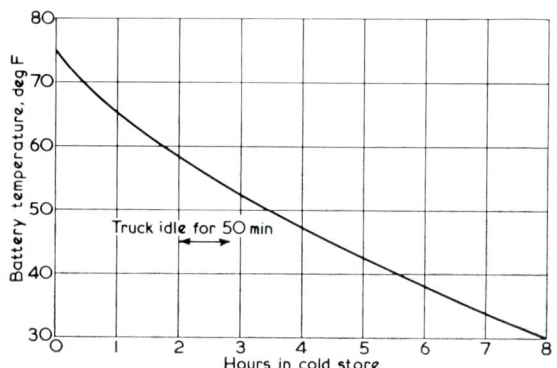

FIG. 5.11. COOLING CURVE OF EXIDE-IRONCLAD "GAUNTLET" BATTERY
OF 18 CELLS, 440 AH

Cold store temperature, $-28·9°$ C ($-20°$ F)

continue to do so on a daily schedule if its temperature was restored to normal during the recharge period.

An indication of the extent of reduction in battery capacity with decreasing temperature of the electrolyte is given in the following table—

Battery (electrolyte) temperature		Capacity available at nominal 5-hr rate
deg C	deg F	per cent
27	80	100
0	32	70
−18	0	40

It should be noted that it is the *electrolyte* temperature which determines the capacity of the battery, and this is affected by several factors besides the obvious one of ambient temperature. These are

(*a*) The temperature of the electrolyte when entering the cold store.

(*b*) The length of time in the cold store.

(*c*) The size and layout of the cells of the battery.

(*d*) The thermal insulation of the battery compartment.

In the interests of maintaining battery capacity, its mean operating temperature should be not less than 0° C (32° F). This is made possible by

(i) Removing the battery from the cold store for recharging, preferably in a room at about 25° C air temperature.

(ii) Ensuring that the temperature of the battery before entering the cold store is 25° C or higher.

(iii) During idle periods such as overnight or at weekends, standing the battery in the warm charging bay, *not* in the cold store.

(iv) Bringing the battery to the fully charged condition on each recharge, and not allowing it to stand discharged in the cold store.

(v) Providing the batteries with a well-fitting lid, as most of the heat losses take place through the cell posts and connectors.

Where batteries are housed in metal trays it is advisable to line the tray walls with heat-insulating material, as this helps to minimize temperature differences between cells in contact with the tray and those in the middle of the battery.

In addition to the reduction in capacity at low temperatures, there are other problems associated with charge acceptance which affect the efficiency of a battery when its temperature is allowed to fall below 0° C (32° F). Difficulty may then be experienced in charging from conventional chargers owing to the high back-e.m.f. of the battery. This is illustrated in the following table—

Battery temperature		Output available at 5-hr rate	Charging voltage per cell at finishing rate		Input required as percentage of nominal 5-hr ampere-hour capacity
			Start	Finish	
deg C	deg F	per cent	V	V	per cent
27	80	100	2·1	2·7	110–115
0	32	70	2·2	2·8	120–125
−18	0	40	2·35	2·95	130–135

With a battery temperature as low as $-18°$ C ($0°$ F), charging from a single taper charger would take about 16 hr. The voltage-time relay would have to be cut out of circuit; otherwise the relay would operate soon after the charge commenced, and the maximum measured time would be the 6 hr on the clock which would be inadequate for charging the cold battery.

FREEZING OF THE ELECTROLYTE

Freezing of the electrolyte is frequently put forward as a likely hazard of low-temperature operation of lead-acid traction batteries, but the facts set out in the table below show that it is very unlikely to occur in practice.

Battery (electrolyte) temperature		Capacity available at 5-hr rate	Electrolyte				Safety margin: difference between actual and freezing temperatures	
			Final sp. gr.	Freezing point				
deg C	deg F	per cent		deg C	deg F		deg C	deg F
27	80	100	1·120	-10	14		37	66
0	32	70	1·170	-19	-2		19	34
-18	0	40	1·216	-33	-28		15	28

There is therefore an inherent protection against the possibility of freezing: the capacity available at a low electrolyte temperature is insufficient to reduce the specific gravity to a value which allows the acid to freeze at that particular temperature.

Some risk of freezing would occur if a battery were fully discharged at normal temperature so that the specific gravity was very low, and the battery were then left in a cold store for a sufficient length of time to cool down to a much lower temperature.

Charging

Operating conditions at most traction battery installations usually allow a charging period of 10 hr or more, and the charger equipment used is fairly simple in design, providing an inherently safe characteristic and automatic means of terminating the charge when the battery is fully charged. This class of charger is known as the *single-step taper charger*.

At installations where special circumstances necessitate recharging in less than 10 hr, it is not feasible to use single-step chargers, and it is then necessary to use a charger with a stepped characteristic. The type most widely used for charging in about 8 hr is the *two-step charger*.

SINGLE-STEP CHARGING

The single-step taper charger is designed to provide a decreasing current as the battery voltage rises on charge. Battery manufacturers recommend that the output of taper chargers should be stated in terms of current at 2·1 and 2·6 V per cell, as these values represent fairly closely the average of the actual voltages obtained at the start and finish of a recharge. The steepness of taper in charging current between these voltage limits is important because it affects the cost of the charger, its sensitivity to mains voltage fluctuations, and the time required to charge.

Two basic tapering characteristics are accepted as being satisfactory: a characteristic representing a 2:1 taper of charging current between the range 2·1 and 2·6 V per cell; and a characteristic representing a taper of approximately 1·7:1 for the same voltage range. This corresponds to 100 per cent charger output at 2·1 V and 50 per cent at 2·6 V for the 2:1 taper characteristic; and 100 and 59 per cent charger output for 2·1 and 2·6 V respectively, for the 1·7:1 taper charger.

For example, a charger designed to give 100 A with a 2:1 taper would give

100 A at 2·1 V

50 A at 2·6 V

or, with a 1·7:1 taper, would give

100 A at 2·1 V

59 A at 2·6 V

Whatever the taper, once the battery reaches gassing point the charger output must not exceed the safe value recommended by the battery maker. This is usually stated in charging rate at 2·5 V per cell, since this represents the average voltage obtained during the gassing stage. This rate, specified by all battery makers and referred to as the *taper rate*, is the current in amperes equal to one-twelfth of the nominal 5-hr capacity, that is 8·33 A per 100 Ah. A typical single-step taper charger characteristic and battery recharge characteristics with this type of charger are shown in Fig. 5.12.

The choice of a suitable taper charger for use with a given vehicle battery depends upon the ampere-hour capacity, and the time available for charging, which should be such that the fully discharged battery could be completely recharged in that time. This period should not normally be less than about 7–8 hr.

Regular recharging of substantially discharged batteries in a shorter period than 7 hr may elevate their average operating temperature above that normally recommended, and temperatures in

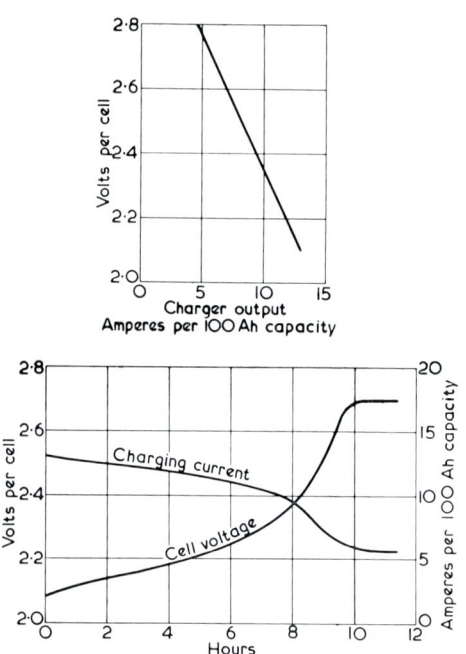

Fig. 5.12. Charger and Battery Characteristics for a Typical
Single-step Taper Charge

excess of 44° C (110° F) may occur at the end of recharge. The rate of heat dissipation of a battery is relatively low, particularly with high-capacity batteries, and persistent high temperatures increase the wear and tear on the battery and shorten its life.

Most charger nomenclatures include reference to the number of cells in the battery, and the charging current at a voltage equivalent to 2·1 V per cell. For example, a designation of 18/75 means that the charger is designed to charge a battery of 18 cells with a starting current of 75 A at the equivalent of 2·1 V per cell.

TWO-STEP TAPER CHARGER

The two-step taper charger recharges a battery in two tapered steps (Fig. 5.13) and provides a fairly high initial rate of charge over the first stage, causing the gassing point to be reached earlier than with a single-stage taper charger having a lower initial output. In the second stage the taper is similar to that of the single-stage

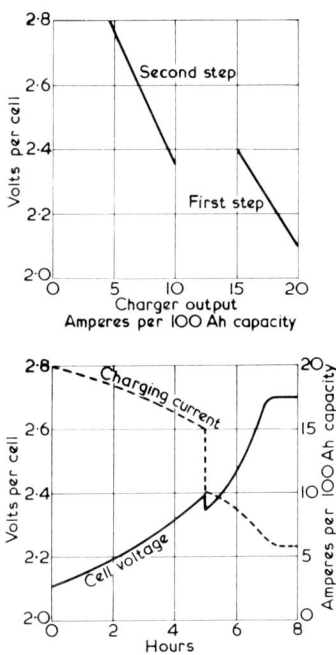

FIG. 5.13. CHARGER AND BATTERY CHARACTERISTICS FOR A TYPICAL TWO-STEP TAPER CHARGE

charger, so that the current at the equivalent of 2·5 V per cell is 8·33 per cent of the battery capacity, and once gassing starts, the outputs of the two chargers are substantially the same.

AUTOMATIC TERMINATION OF CHARGE

The amount of charge is controlled automatically by means of a voltage relay and time switch. The voltage relay, set to operate between 2·35 and 2·4 V per cell, starts the timing motor, which runs for a pre-set time before switching off the charger. The principle of the voltage-timing relay is shown in Fig. 5.14. It will be seen that the timed portion of the charge is virtually the same for the battery in different states of discharge, and the variable period occurs in the first stage up to the time the voltage relay operates. In the range of current available from single- and two-step taper chargers, the amount of charge put into a discharged battery by the time its voltage reaches 2·35 to 2·4 V per cell is about 75 per cent of that required to give a full charge. Replacing the remaining 25 per cent requires a

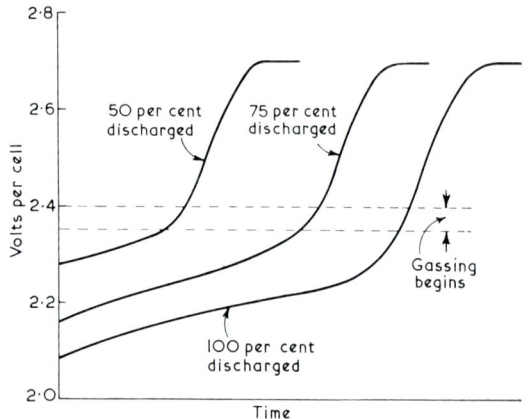

FIG. 5.14. CELL VOLTAGES ON CHARGE SHOWING CONSTANT GASSING PERIOD

time of from 3 to 4 hr, depending on the charging current and its rate of fall with rise in battery voltage. The setting of 2·35–2·4 V is chosen because a battery will always attain this value under normal charging conditions.

It is sometimes suggested that the setting should be adjusted to some value representing the fully charged voltage of the battery, but this would not be feasible, since this voltage varies with the age of the battery, its temperature, and the charging current.

CURRENT-VOLTAGE CONTROLLED CHARGERS

The class of charger using semiconductor rectifiers and transistor control of voltage and current output is coming on the market to challenge the universally used taper chargers described above. The "controlled" chargers are capable of providing more flexible ways of charging at about the same cost as the conventional charger. The output of the charger is controlled at virtually constant current below a preselected voltage (Fig. 5.15). When this voltage is reached the system is maintained on constant voltage and the current gradually tapers.

The selected voltage for traction battery charging must be set at a safe level so that there is no excessive gassing under all conditions of battery and service operation. A hot, or old, battery accepts more current at constant voltage than a new, or cool, battery. This is one of the difficulties in charging traction batteries in parallel from this type of charger. For normal operation the selected constant voltage level has to be not more than about the equivalent of 2·35 V per cell,

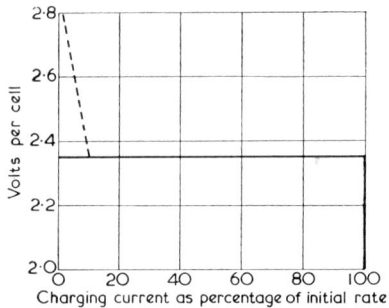

FIG. 5.15. TYPICAL CHARACTERISTIC OF CURRENT-VOLTAGE
CONTROLLED CHARGERS

and for traction work this setting is not high enough to give complete charging. Frequent equalizing charges must therefore be given.

Some chargers incorporate a voltage uplift characteristic (Fig. 5.15, broken line) once the current has fallen to a low value, to bring the voltage above 2·35 V so that some gassing takes place at the end of the charge, but it is still necessary to give equalizing charges at suitable intervals.

Note. It is advisable to consult the battery manufacturer before installing any traction battery charger equipment other than the inherently safe single- or two-step taper chargers.

EQUALIZING CHARGES

To ensure that every cell of a battery is maintained in a healthy condition so that maximum life is obtained, it is advisable to give an extended charge every two to three weeks. This extra charge should be given at the low "equalizing" rate shown in battery makers' catalogues, or at a rate between 2 and 5 A per 100 Ah capacity.

The equalizing charge is usually given for a period of at least 12 hr or until every cell is gassing freely, and readings of voltage and specific gravity have ceased to rise. Where it is not possible to give regular equalizing charges, the routine charge must be adjusted so that over a period of several weeks there is no falling off in the specific gravity of any cell. This means that the amount of daily charge is more than would be given where regular equalizing charges are possible. When equalizing charges are given it is possible to terminate daily routine charges when the specific gravity is within 10 points of that reached on equalizing charge.

If at all possible, cell readings of voltage, specific gravity and temperature should be recorded towards the end of the equalizing charge. These readings serve as a guide to the state of charge of the battery and are a useful check of any change in cell condition during the life of the battery.

No special skill is required in taking cell readings, and a little practice in using the hydrometer, or voltmeter, is all that is required of the average operator.

It should be noted that specific gravity readings on charge are unreliable unless taken at the very end, or in the last half hour, of the charge, and with the electrolyte at the correct level at the start of charge. Low levels in service usually result from loss of water during gassing on charge, and the specific gravity of the electrolyte with the battery fully charged will be higher than normal.

Tapped batteries require special attention as several cells are tapped for circuits working at a voltage lower than that of the battery. (For example, a 36-V battery may be tapped for auxiliary 12-V circuits, such as horn, lights, or radio control equipment.) This results in the tapped cells becoming discharged further than the remaining cells, but if the tapped cells are kept charged, the remaining cells will be overcharged, which may affect their life.

If possible, therefore, taps on batteries should be avoided by using equipment rated for the full battery voltage, or providing a dropping resistor in series with the auxiliaries.

Charging Stations

In large works where many battery trucks may be operated over an extensive area, it is advisable to provide more than one charging station to avoid long runs to and from the working areas.

The space to be provided in the charging station will depend on the number of batteries to be charged at one time. The batteries on industrial trucks which work for a fixed period of about 8 hr each day will not need changing, and will be charged in position on the trucks, Fig. 5.16. Space must therefore be provided for the trucks, with their respective chargers mounted on the walls or suspended on frameworks in conveniently spaced rows.

FIG. 5.16. CHARGER INSTALLATION FOR FORK-LIFT TRUCKS

Batteries not removed from trucks for charging

(*Westinghouse Brake & Signal Co. Ltd.*)

FIG. 5.17. CHARGER INSTALLATION FOR FORK-LIFT TRUCKS ON SHIFT DUTY

Batteries changed at set intervals and charged off the truck

(*Westinghouse Brake & Signal Co. Ltd.*)

At large installations where shift work is operated and trucks are working almost continuously, it is necessary to provide spare batteries which are changed at set intervals. In this case it is usual to provide space for the batteries alone, which may be arranged along the walls and in rows across the floor (Fig. 5.17).

General recommendations which apply to all charging stations are—

1. The station should be well ventilated, reasonably cool, and spacious so that trucks and batteries may be maintained in the same room.

2. Overhead lifting tackle or a fork-lift truck with lifting attachment should be available for lifting batteries out of trucks. If battery changing takes place regularly these facilities are essential.

3. A supply of running water for cleaning purposes, and a supply of pure, approved water for topping-up batteries, should be available. The pure water should be stored in a clean carboy or lead-lined tank with top cover.

4. It is advisable to provide instruments for checking battery condition; these may consist simply of a hydrometer and thermometer. A reliable cell-testing voltmeter is also useful.

5. A first-aid cabinet containing an eyewash bottle should be fixed in a prominent position. Acid splashes in the eye require prompt attention, and bottles are designed so that the injured person can flush the solution into his eye without delay. Medical attention should always be sought as soon as possible.

6. An instruction card describing first-aid treatment for electric shock should be read by all personnel using charging equipment. The card should be fixed in a prominent position in the charging station.

7. More than one operator should be in attendance at a time when a charging voltage higher than 110 V is used.

8. A NO SMOKING notice should be prominently displayed.

CHARGING STATION LAYOUT

It is impossible to set out precise recommendations for station layout unless information relating to the number of vehicles and space available is provided.

In moderately sized stations the chargers can be mounted conveniently on walls or supporting frames (Figs. 5.16 and 5.17), in such a way that there is ease of access and the trucks or vehicles can be run in as close as possible to the chargers or spare batteries.

The charger should be mounted clear of the floor at a height convenient for the operator to work the controls. This will also prevent the charger getting wet when floors are washed down.

A clear space for air ventilation of about 18 in. on all sides of the charger should be allowed, and the free flow of cooling air must not be restricted by placing material over the charger. It should be installed away from radiators or other sources of heat.

Each truck or vehicle should be assigned its own particular charger —this is imperative where truck batteries vary in voltage and capacity. Batteries should be correctly paired with the right size of charger, and some simple method of identification should be devised.

Where the charging load is considerable and chargers are connected to the same supply, it is advisable to distribute the load over a 3-phase supply.

The supply leads to each charger should be provided with fuses and control switch, to enable work to be carried out on any charger without interfering with the operation of the remainder.

In estimating mains cable, plug, switch and fuse sizes, the following figures may serve as a rough guide.

Example

Charger rating (amperes, d.c.) for any recharge period between 10 and 14 hr

$$= \frac{1\cdot43 \times \text{Ah capacity of battery at 5-hr rate}}{\text{Charge period in hours}}$$

Thus, for an 18-cell 702-Ah battery to be charged in 12 hr the charger rating would be

$$\frac{1\cdot43 \times 702}{12} = 84 \text{ A}$$

The nearest available taper charger would have a rating of 85 A and either a single-phase or a 3-phase charger could be used.

Single Phase. The maximum value of initial alternating current is given by

$$\frac{4 \times \text{No. of cells} \times \text{Initial direct current}}{\text{A.C. voltage per phase}}$$

For the 18-cell 85-A charger with a 240-V a.c. supply this becomes

$$\frac{4 \times 18 \times 85}{240} = 25\cdot5 \text{ A}$$

Three Phase.

$$\text{Current per phase} = \frac{2 \times \text{No. of cells} \times \text{Direct current}}{\text{A.C. voltage}}$$

For same charger connected to a 415-V 3-phase supply,

$$\text{Alternating current per phase} = \frac{2 \times 18 \times 85}{415} = 7 \cdot 4 \text{ A}$$

CHARGING ROUTINE

In many charging stations, battery charging usually takes place between the hours of 6 p.m. and 6 a.m. to take advantage of cheap off-peak tariffs. In such stations, chargers would be switched on and off together by means of a central time control. Alternatively, charging would be automatically terminated by the operation of the voltage-time relay of the charger.

At other charging stations, batteries are put on charge at the end of the working day. No special skill is required for routine battery charging using taper chargers with automatic control. The charging station attendant or the driver of the truck starts the charge by switching on the a.c. supply and resetting the voltage-time relay.

Charging takes place overnight and is terminated automatically when the battery is charged. The following day the driver on collecting his truck, with the battery in a fully charged condition, switches off the a.c. supply, disconnects the battery from the charger and connects it to the plug on the truck.

Where battery trucks are operated continuously it is necessary to use spare batteries, housed in the charging area, which are changed at about 8-hourly intervals. In theory it is possible to work a 24-hr schedule using two batteries and one charger per truck, where the charger is designed to recharge a battery in $7\frac{1}{2}$–8-hr. Under these conditions, batteries are alternately discharged and charged without any rest period. In practice it is advisable to use three batteries and one 8-hr charger per truck, which enables each battery to have an 8-hr rest period in every 24 hr. If three batteries (Nos. 1, 2 and 3) are being discharged in 8 hr and recharged in 8 hr or slightly less, the schedule becomes—

Eight-hour periods	*Discharge*	*Charge*	*Rest*
1st period (A)	Battery 1		
2nd period (B)	Battery 2	Battery 1	
3rd period (C)	Battery 3	Battery 2	Battery 1
4th period (A1)	Battery 1	Battery 3	Battery 2
5th period (B1)	Battery 2	Battery 1	Battery 3

and so on.

The rest period for each battery is important in restricting battery operating temperature to a reasonable level (below 43° C (110° F)) and so helping to prolong battery life.

Battery Maintenance

CARE OF BATTERY IN SERVICE

1. Keep all naked flames away from the battery, particularly during and shortly after a charge. This is to avoid the risk of an explosion caused by ignition of the gaseous mixture in the top of the cell and immediate vicinity.

2. Keep battery compartments open by removing any top cover during a charge.

3. Keep all cells, trays and containers dry and clean. Any corrosion of metalwork should be removed, by scraping and neutralizing any acid with a solution of dilute ammonia or a solution of sodium carbonate (washing soda). Protect from further corrosion by covering with acid-proof paint or petroleum jelly. Wash and dry the trays occasionally and apply a coat of acid-proof paint. Smear all metal intercell and terminal connectors with petroleum jelly to protect against corrosion. Do not use oil or grease as they will damage cell containers and sealing compound.

4. Keep all bolted connexions clean and tight. Check the plugs and sockets used for connecting the battery to the charger, or the truck, for overheating due to indifferent contacts.

5. Add nothing but pure water to the cells. This should be done at about weekly intervals.

6. Top up to the correct level, or if in doubt undertop, to avoid acid overflowing from the cells when gassing on charge.

7. Never add acid to the cells unless it is known that acid has been spilt, and even then it is advisable to consult the battery maker first.

8. Do not use metal containers when topping up, because of the danger of short-circuiting the live terminals and connectors.

9. Wipe clean and dry the cell lids after topping up, and replace vent plugs, which should be securely tightened in position at all times. They should be removed only for adding water or taking hydrometer readings.

CARE OF BATTERY OUT OF SERVICE

If a battery is taken out of service for a time, or if a new charged battery cannot be put into service immediately, it should be given an equalizing charge, wiped clean, and stored in a cool, dry place. All detachable connectors should be disconnected to reduce electrical leakage. Every month check the acid level and give an equalizing charge.

If a vehicle is used at irregular intervals the battery should be given an equalizing charge every month.

In special circumstances where the battery is stored for periods not exceeding 6 months, it can be maintained on trickle charge. The charging current in milliamperes is usually between one and three times the nominal 5-hr ampere-hour capacity. The ideal current is that which will maintain specific gravities around the fully charged value, without any appreciable gassing.

Batteries supplied in a dry, uncharged condition should be stored in a cool, dry place with the vent plugs securely in position. It is recommended that under adverse storage conditions the battery should be filled and charged within 2 years from the date of dispatch.

PUTTING BACK INTO SERVICE

For charged batteries, check the levels and give an equalizing charge, before putting back into service and again after the first discharge, in order to bring them into good working condition.

For dry and drained batteries, fill with pure sulphuric acid of the correct density, and charge in accordance with the battery makers' instructions.

Operating Costs

It is recognized that the operating costs of battery-driven vehicles are somewhat lower than those of petrol- or Diesel-driven vehicles. It is rather difficult to make an accurate comparison of total operating costs of the various types of vehicle unless payload capacity and operating conditions are exactly similar. However, reliable comparison figures do show a difference in total operating cost of several pence per mile in favour of a battery-driven vehicle compared with a petrol vehicle.

Almost the whole of this saving is due to the lower running costs of the battery vehicle as a result of lower fuel (electricity) charges. Maintenance costs for the battery vehicle would also be somewhat lower.

COST OF RECHARGING

A rough estimate of the current consumed in fully recharging a traction battery which has been discharged to the extent of a routine discharge amounting to 85–90 per cent of its nominal 5-hr capacity may be calculated as follows—

$$\left.\begin{array}{l} \text{Number of units (kWh)} \\ \text{required to charge} \end{array}\right\} = \left\{\begin{array}{l} 1\cdot9 \times \text{Nominal kWh} \\ \text{capacity at 5-hr rate} \end{array}\right.$$

the nominal Wh capacity being the Ah capacity multiplied by the nominal battery voltage.

Example

Find the number of units required to charge an 18-cell 702-Ah battery.

$$\text{Nominal kWh capacity} = \frac{18 \times 2 \times 702}{1,000}$$

$$\text{No. of units required} = \frac{1\cdot9 \times 18 \times 2 \times 702}{1,000} = 48 \text{ (kWh)}$$

With electricity at 1d. per unit, the cost would be 4s.

An alternative method would be to calculate the watt-hour input to the battery using an average charging voltage of 2·35 V per cell, and taking the Ah input as 115 per cent of the output. With the battery discharged to 85 per cent capacity (600 Ah),

$$\text{Energy input} = 18 \times 2\cdot35 \times 600 \times 1\cdot15 \text{ Wh}$$

With a charger efficiency of 60 per cent,

$$\text{Energy consumed} = \frac{18 \times 2\cdot35 \times 600 \times 1\cdot15}{1,000 \times 0\cdot6} = 48\cdot5 \text{ kWh}$$

A comparison of charging costs based on current at 1d. per unit for various sizes of traction battery is made below.

Application	Battery		Energy consumed	Cost of electricity	
	No. of cells	Cap-acity			
		Ah	kWh	s.	d.
1000-kg road vehicle .	36	219	30	2	6
1500-kg fork-lift truck .	12	611	28	2	4
14,000-kg mining locomotive .	100	504	191	15	11

CHAPTER 6

CAR AND HEAVY-VEHICLE (AUTOMOTIVE) BATTERIES

FOR the purposes of this chapter, the term "heavy vehicles" will include all commercial vehicles, buses, lorries, tractors, earth-moving equipment, etc.

PROSPECTIVE MARKET

Motor vehicles of all kinds are being produced in ever increasing numbers all over the world. In the United Kingdom, production is such that it amounts to almost three cars and one commercial vehicle for every minute of every day. All these new vehicles, whatever their size and purpose, together with the millions already in service, require a battery to supply electrical power for engine starting, lighting and other duties.

The demand for lead-acid automotive batteries is therefore two-fold: firstly, as initial equipment on new vehicles; secondly, as replacement units on a life expectation of approximately two to four years. With this almost insatiable demand it is not surprising, therefore, that the manufacture of automotive batteries consumes more than 80 per cent of the lead used in the battery industry.

PROTOTYPE CAR BATTERIES

In the pioneer days of the motor-car, or horseless carriage, where coil ignition was used, dry primary batteries consisting of three or four cells in series supplied the current. Because of their relatively short life and uncertain performance, a cell-testing voltmeter was usually a necessary part of the kit housed in the generous sized tool box carried on the car's running board. It was essential to check the condition of these dry cells at frequent intervals, and a spare battery was just as necessary as the spare can of petrol or spare wheel when a journey of any distance was contemplated.

Engines were started mainly by hand cranking, and apart from the coil ignition there was no other electrical component requiring a battery. Lighting when used was by means of oil or acetylene.

INTRODUCTION OF LEAD-ACID BATTERIES

With the rapid development of the lead-acid battery in the early 1900s, there was a steady change from dry cells to this type of battery for coil ignition on motor-cars. Batteries consisted of two or three lead-acid cells, usually in celluloid boxes housed in an outer wood container mounted on the running board. As yet there were no car-type generators and discharged batteries had to be removed and recharged at a charging station. Batteries were rated by the mile, as, for example, 300, 400 or 500 miles, which represented the distance over which they could supply the ignition current before requiring a recharge.

ELECTRIC STARTING AND LIGHTING

Considerable thought and ingenuity were continually being applied to improvements in motor-car design, including systems for the automatic starting of engines, and in 1911 a major success was achieved when the electric self-starter was invented in America. This gave a tremendous stimulus to car battery manufacture, for there was a demand, not only for more lead-acid batteries but also for batteries of much greater capacity to supply the considerable power required to drive the electric starter motor for turning the engine.

At the same time, electric lighting was being introduced into some offices, public places, and the larger stately homes, and it was only natural that it should soon have been extended for use on cars. This again was a great improvement on the old messy systems of lighting by oil or acetylene, which entailed constant care and attention. For the first time, batteries were now supplying electrical power for starting, lighting and ignition, and as a result of this triple duty they became to be known as "S.L.I. batteries." The increased demand on the battery in supplying these loads made it imperative for some means of charging to be carried on the car, otherwise the battery would quickly have become discharged. The problem was solved by adding a generator driven from the engine, which provided charging current for the battery whilst the car was being used.

BATTERY TYPES AND SIZES

The early lead-acid car batteries did not conform to any standard voltage or design, varying between 6 V and 24 V (3-cell and 12-cell batteries) with a preference for 6 V as electrical systems became more widely used. This 6-V system utilizing a 3-cell lead-acid battery was popular for some twenty years following the introduction of the electric self-starter.

During that period, considerable improvements were made in all the electrical components, including the battery. The original battery, consisting of three cells in individual celluloid boxes, gave way to cells assembled in ebonite boxes contained in a polished wood box complete with lid. This shared pride of place on the running board with the tool box and spare wheel. As running boards began to shrink, batteries disappeared from view and were mounted in the most inaccessible places, including beneath the floor boards and passenger seats. It was not surprising, therefore, that they were very much neglected as even the simple and necessary operation of keeping them clean and topped up with water entailed a major upheaval of the car structure.

The early batteries failed prematurely for a variety of reasons besides that of poor maintenance. Active material or paste was lost from the plate grids, which were inadequately designed to hold the paste under the vibration experienced on the early cars. Short-circuits developed between the plates due to breakdown of the perforated celluloid or ebonite separators, caused by oxidation of the separator material by contact with the lead dioxide of the positive plate.

The introduction of separators made from wood, the best of which was Port Orford cedar, was a definite advance on the old celluloid or ebonite separators, but even the wood separator became perforated by oxidation or charred by the sulphuric acid after a period in service. Thus, even with improvements in grid designs and paste techniques and the use of improved cell containers, the factor which limited the life of the battery was almost always the breakdown of the separator.

CHANGES IN VOLTAGE

Whilst improvements to the battery design were taking place, other changes affecting the voltage of the electrical system were also made. Up to about 1930 most car manufacturers in England and America were using 6-V systems. It was about this time that manufacturers here started to adopt 12-V systems using 6-cell lead-acid batteries, and by the late 1940s this system was the standard for this country. In America, however, no great change from 6-V to 12-V batteries occurred until 1953–54, when most U.S. motor manufacturers adopted 12-V systems. The main points in favour of the 12-V system are listed below.

1. For the same power requirements (volts multiplied by amperes) the current demand on the 12-V system is almost halved. This is an important reduction of the heavy current drain on the battery during engine starting.
2. Because of economy of current with the 12-V system, voltage

drops (amperes multiplied by ohms) in cables and brushes of the starter motor are much lower and the cable size can be reduced.

3. Distributors and voltage regulators are more efficient at the higher voltage because of the reduced current which they have to make and break.

4. The 12-V generator can be smaller compared with the 6-V generator for the same output.

5. Any deterioration in battery condition, even if confined to one cell, can more readily be tolerated in a 6-cell than a 3-cell battery.

6. The 12-V battery is usually more expensive and heavier than the 6-V battery, but for the reason given in the last paragraph, the life of the 12-V battery should be longer than that of the 6-V battery.

Choice of Earthed Polarity

Ever since batteries have been used on motor vehicles there has been a divergence of opinion regarding the earthing, or grounding, of the electrical system. In America and this country both positive and negative earthing have been used at some time or other for motor-vehicle electrical systems. In America, most car manufacturers were earthing the positive pole of the battery to the car chassis until the advent of the 12-V battery about 1953, when there was a complete change to negative pole earthing. British manufacturers were divided between positive and negative earthing until about 1940, when positive earthing became standard practice.

Whichever pole of the battery is earthed, the electrical accessories use single-pole wiring with one terminal of the battery and one side of all other electrical components securely connected to the chassis. The chassis therefore acts as an "earth" or return path for the electric circuit. The advantages claimed for the respective earthing systems are worth noting—

(*a*) Less corrosion of the battery terminal connectors is a claim made for both systems.

(*b*) With positive earthing, ignition is better.

(*c*) With positive earthing, burning or pitting of the electrodes of the plugs and distributor arm is less.

With regard to (*a*), it is a fact that corrosion of battery-terminal cable connectors, which are usually lead-plated brass, will occur in the presence of sulphuric acid with either positive or negative earthing unless attention is given to cleaning them periodically and applying a protective coating of petroleum jelly against electrolytic attack.

Although it is generally agreed that the space between the plug points is more readily ionized and therefore a better conducting path for a spark when the central electrode is made negative (as in positive earthing), the same result can be achieved with negative earthing by reversing the connexions to the coil. It would therefore appear that there are no outstanding advantages which would make either choice greatly superior to the other.

Demands of the Modern Car

The modern car, with its complexity of electrical components and high-compression engine, needs a battery which can supply current for all these accessories and still have power to spare for engine starting. A list of the electrical loads for a typical family car of engine capacity of about 1,500 cc is shown in the following table—

Typical Car Accessories and Ratings

System, 12 V
Battery, 12 V, 40 Ah

Current loads	A
Ignition	2
Side, rear and number plate .	2·5
Headlamps	6·5
Windscreen-wiper motor .	3·5
Heater motor . . .	4
Direction-indicator lamps .	3
Starter motor	
Engine cold . . . 115–130	
Engine warm . . . 90–100	

When the battery is receiving no charge from the generator, either because the engine has stopped, or because the generator is running below cut-in speed, it might have to supply a load of about 19 A, or if parked with only side and rear lights on, about 2·5 A. To provide 19 A for a period of approximately one hour and then to be capable of supplying 120 A at 9 V (about 1,100 W) for at least 10 sec for engine starting, would require a 12-V battery of 37-Ah capacity. In fact, 12-V batteries varying in capacity from about 40 to 50 Ah are usually fitted to most family-size cars, and a battery of the smallest capacity would therefore be capable of starting the engine even when previously discharged by about 50 per cent of its capacity.

Under most operating conditions, car batteries are seldom more than about 25 per cent discharged, which is equivalent to 4-hr parking with lights on, so that the 40-Ah battery usually has a margin of capacity for engine starting and accessory loads. Under normal circumstances, this margin is very desirable as the demand on the battery greatly increases under adverse conditions of low temperature or poor engine response. The capacity margin also compensates for reduction in battery capacity with age.

LOW-TEMPERATURE OPERATION

In this country, the temperature of automotive batteries rarely falls much below about −5° C (25° F), and then for not more than a few weeks during an average winter. However, because many of our cars are exported to Canada and North America, where winter temperatures can be extremely low, the battery must be capable of supplying power for engine starting at much lower temperatures. American, Canadian and, more recently, British battery specifications include a low-temperature clause which requires the battery to supply a hundred or more amperes, depending on battery size, for several minutes at a battery temperature of −18° C (0° F). Good low-temperature performance is therefore a very important feature of the automotive battery, and is also a sure criterion of a quality battery. It is true to say that only those manufacturers who are skilled in battery techniques can supply batteries which consistently and repeatedly meet starting requirements at −18° C (0° F).

Battery Sizes and Construction

The sizes of automotive battery most commonly used for private cars are the 12-V 7-plate and 9-plate. These are the numbers of plates per cell, so that the total number of plates in each 6-cell battery is 42 or 54.

The 6-V battery is not now widely used and is mainly confined to cars made abroad. Since it has only half the voltage of the 12-V battery, it must have a greater ampere-hour capacity to supply the same power. This entails more plates per cell, and the most popular sizes have 13, 15, or 17 plates, giving a total of 39, 45, or 51 plates respectively for the 3-cell battery.

PLATES, GROUPS AND ELEMENTS

The plates used in the automotive battery are of varying dimensions but usually about 13 cm square and just less than 2·5 mm thick. Normally the positive plates are 1·8–2·5 mm thick, and the negative plates 1·5–1·8 mm thick. A 9-plate cell consists of 4 positive and 5 negative plates. The 4 positive plates, suitably spaced apart, are connected together by burning a lead bar to the plate lugs. The lead plate bar also carries the terminal post, and the whole assembly is called the "positive plate group." Similarly the 5 negative plates are burned to a common plate bar with terminal post to form the negative plate group.

The two groups are interleaved, and the plates are insulated from each other by inserting the separators (8 in number) between the plates, making an assembly known as an element (Fig. 2.3). Six such elements are assembled in a 6-compartment hard-rubber container, and each element is fitted with a lid which is sealed to the walls of the container by sealing compound. The cells are then connected together by burning antimonial-lead intercell connectors on to the opposite poles of adjacent cells.

The battery is completed by building the end terminal taper posts and fitting the vent plugs to the cell lids. The taper posts are distinguished by the letters P and N or the signs + (plus) and − (minus), but as an extra precaution the positive post is made of greater diameter than the negative to prevent the cable clamp connectors being wrongly connected.

Battery Design

The performance of lead-acid batteries in terms of electrical output and life has steadily advanced over the last 20 years, owing to improvements in design involving almost all the various component parts.

PLATE GRIDS

The grids (Fig. 2.1), which hold the paste, or active material, and serve as electrical conductors for the current, were for many years made in an alloy consisting solely of lead and antimony in the approximate proportions of 90 per cent lead to 10 per cent antimony. The antimony was added to produce better, sharper and stronger castings, and the antimonial-lead grid had a greater resistance to anodic corrosion than a pure lead grid.

The addition of antimony in these proportions produced some disadvantages resulting from the antimony being transferred from the positive grid to the negative plate as the positive grid corroded in service. The poisoning of the negative paste by the antimony results in loss of charge due to chemical action between the lead and the antimony, which becomes worse as the batteries age.

This trouble has been largely overcome by reducing the antimony content of the lead alloy, whilst improving the resistance of the grid of the positive plate to anodic corrosion, by adding small quantities of such metals as silver, copper, tin and arsenic. The success of these new alloys in retarding corrosion has enabled grids thinner in section and lighter in weight to be used.

PASTES

The oxides of lead, which are mixed with dilute sulphuric acid to form the pastes which are converted electrochemically to the active materials of lead dioxide and lead, are chosen to provide maximum capacity compatible with long life. This has been achieved by the use of finely ground oxides and improved control of mixing and setting techniques.

Vast improvements have been made in low-temperature performance by the use of new ligneous organic compounds, or expanders, in the negative paste. These maintain the lead particles of the negative paste in the spongy form, which is essential for providing high starting power in low-temperature operation.

SEPARATORS

It was essential that improvements in separator design and quality should keep pace with improved grids and pastes. The extent of the improvements in the separator has been such that most of the modern separators outlive the plates and no longer determine the life of the battery.

Some of the best separators are made from microporous rubber or microporous plastic. One such separator is made from polyvinyl chloride (p.v.c.) and is ideally suited for automotive batteries. It is 80 per cent porous, has low electrical resistance, and is extremely resistant to oxidation and inert to sulphuric acid.

Fibre-based separators also are widely used. These are made from specially selected paper strengthened by a resin treatment. The porosity and resistance characteristics are good, but because under adverse conditions they become perforated owing to oxidation, they do not have the long life of the microporous plastic or rubber separators.

SEPARATOR RESISTANCE

A feature of all modern separators which distinguishes them from the older types is their low electrical resistance. The resistance of a typical separator is of the order of $0.26\,\Omega$ per cm^2 at a temperature of $21°\,C$ ($70°\,F$), and its total resistance is $0.0015\,\Omega$.

CONTAINERS AND LIDS

Modern containers are lighter and stronger than the earlier composition types. They are moulded multicell monobloc units of 3 or 6 cells, using hard rubber or resin rubber materials. They are designed to withstand considerable vibration and rough handling under widely differing temperatures, and are resistant to acid attack or penetration.

Lids may be made one for each cell, or as a single one-piece lid, or monolid, which is sealed in position above the cells.

FIG. 6.1. MODERN CAR BATTERY WITH ONE-PIECE
LID MONOVENT COVER, AND "CORRECT
LEVEL" DEVICE

One of the latest types of car battery using a polypropylene container and cover is shown in Fig. 6.1.

The introduction of these car batteries in bright coloured plastic containers gives them a completely new look and removes the old black box image of conventional types.

Other important features incorporated in this latest battery by Exide include the use of shortpath, through-partition, intercell connectors. These, together with the lightweight container and the use of new (patented) active material in the plates, provide considerably more starting power per unit weight and volume of battery than was possible before. The battery is fitted with an automatic one-shot filling/levelling device which ensures a speedy and accurate method of topping up. The levelling device relies on a ball valve which is sealed by the ball with the top cover removed for topping-up purposes. When water is poured into the lid trough, it flows down the filling tubes into the cells. An air lock is created inside the cell when the levels inside reach the bottom of the filling tubes. The level in each cell can then rise no further. As more water is added this fills each tube and overflows to fill the trough.

Pouring should stop when water begins to collect in the bottom of the trough. Replacing the top cover displaces the balls from the valve seating, so breaking the air lock. The water in the trough and tubes then flows into the cells, filling each one to the correct level.

INTERCELL CONNECTORS

The intercell connectors are antimonial-lead straps welded across adjacent cell posts so as to conduct the current through each cell.

All battery connectors are designed to carry safely, and without undue voltage drop, the one-minute current, which for the 40-Ah battery is 280 A. This gives a generous safety margin for the duty the battery is designed to perform. Fused connectors are usually due to high-resistance contact resulting from bad welding during manufacture, or to the connectors passing excessively high currents owing to a short-circuit across the battery terminals.

THE ELECTROLYTE

The electrolyte is the only component which has remained unchanged during a century of battery manufacture. The choice of acid concentration (specific gravity) may vary with the type of battery, the operating temperature, or with different battery makers' preferences, but without exception the electrolyte is pure dilute sulphuric acid. In temperate climates with air temperatures below 32° C (90° F), the specific gravity of the sulphuric acid in fully charged batteries varies between 1·270 and 1·285 corrected to 15·6° C (60° F). In tropical climates with air temperatures above 32° C (90° F) the specific gravity varies between 1·230 and 1·250 (corrected to 15·6° C).

Dope Electrolytes or Battery Additives

Periodically much publicity is given to special additives for car and motor-vehicle batteries. Spectacular and extravagant claims are made for these additives for extending battery life or reviving old and ailing batteries. Treatment usually consists in adding the patent "elixir" to each cell of the battery followed by a long charge at about half the normal charging rate. The beneficial effects of this treatment on batteries which have become discharged or sluggish in service are frequently immediate and noticeable when the battery is put back on the car and the starter motor operated. But similar results can be achieved by simply giving a long charge without adding "dope" to the battery. The large battery manufacturers spend vast sums of money in research for materials or methods of manufacture which will extend battery life or improve performance, but so far no electrolyte additive of any great merit has been invented. It is also of interest to note that most battery makers declare their guarantees void if anything other than pure water or sulphuric acid is added to batteries of their manufacture.

Putting into Service

Before a battery can store or deliver electrical energy it must be filled with dilute sulphuric acid and in most cases given a charge. There are, however, dry-charged batteries which can be put into service after filling with acid without any preliminary charge. The plates in these batteries are carefully dried after receiving a formation charge, and particular care is taken with the negative plates by quick drying in an inert atmosphere such as hydrogen or nitrogen to ensure that there is negligible oxidation during the drying process. The maximum storage period for these batteries is approximately $1\frac{1}{2}$ to 2 years, after which they require a charge before being put into service because of the gradual loss of dry charge characteristics due mainly to oxidation of the negative plates.

Batteries can be grouped into three classes as regards requirements for storage and putting into service, namely uncharged, dry-charged, and filled and charged.

UNCHARGED BATTERIES

Batteries sold as uncharged contain dry plates and dry separators, and are capable of being stored indefinitely without harm. They are prepared for service by filling with pure sulphuric acid usually of 1·260 sp. gr. After a 6–12-hr stand period to allow the acid to soak into the plates, charging is recommended at a current equal to about one-tenth of the rated 20-hr capacity. The charge continues for at least 48 hr. At the conclusion of the charge, when voltage and acid density readings are constant, the electrolyte is adjusted to the value specified by the battery maker.

Dry-charged Batteries

These contain specially processed, or dry charged, plates and dry separators, and usually the vent plugs are taped over to restrict "breathing" within the cells. The batteries are capable of being stored dry for the period claimed by the manufacturer. They are prepared for service by filling with acid of 1·270–1·280 sp. gr. at a temperature of 21°–27° C (70°–80° F). After a stand period of one hour the electrolyte level in each cell is readjusted, and the battery can be put into service when it should be capable of supplying 80 per cent of its rated capacity. If the battery is stored beyond the expiry date shown on the label, it must receive a charge, usually for 12 hr, before being put into service.

Filled and Charged Batteries

These are capable of either being put into service immediately or being stored for long periods. A condition of storage is that they receive periodic freshening charges to replace the loss of charge due to internal chemical action. The frequency of the freshening charge should be about once every 2 months at a current equal to one-tenth of the 20-hr capacity, until specific gravity and voltage readings are constant. If a battery is required for service at any time between freshening charges, the specific gravity of its electrolyte should be checked. If this is more than 10 points lower than the fully charged value, the battery should be given a freshening charge before being released.

Capacity Ratings

Automotive batteries are given a capacity rating at the 20-hr rate, expressed in ampere-hours. At the same time, since the capacity of a battery is directly related to rate of discharge, battery temperature and final voltage, all these factors should be stated when quoting the capacity. For example, a 40-Ah battery can supply 2 A continuously for 20 hr at a temperature of 25° C (77° F) to a final voltage of 10·5 V. Similarly a 100-Ah battery is capable of supplying 5 A continuously for 20 hr at 25° C (77° F) to 10·5 V.

At one time the automotive battery was rated at the 20-min rate as well as the 20-hr, but neither has much practical significance when related to actual service on the motor vehicle, where the battery is providing, intermittently, a wide range of current. This can vary from several amperes for ignition purposes to a hundred or more amperes for starting.

Nevertheless, in all automotive battery specifications, the 20-hr rating test is standard and serves as a useful guide to battery design. The 20-hr test exhausts the battery to its practical limit of capacity and therefore provides an indication of the amount of active material available and also the balance of active material to acid. Most automotive batteries are designed to have a final acid specific gravity of 1·100–1·130 at the completion of discharge at the 20-hr rate.

Standards for Motor Vehicle Batteries

For many years the automobile industries in most countries have applied the principles of standardization to their products in the interests of efficiency and economy.

In the U.S.A., the Society of Automotive Engineers (S.A.E.) first published Standards for batteries in 1914, with additions and revisions right up to the present time. In the U.K., the British Standards Institution produced in 1965 British Standard 3911: "Specification for Starter Batteries for Internal Combustion Engines". Six tests are listed which must be met before a battery can be accepted as having the required quality and performance. These are—

1. Rating test at the 20-hr rate at 25° C (77° F)
2. High-rate discharge test at 25° C (77° F)
3. High-rate discharge test at −18° C (0° F)
4. Test for retention of charge
5. Life cycling test
6. Overcharge test

For laboratory tests to have any practical value or significance they should simulate, as closely as possible, actual service conditions. This is not easy to accomplish as no two batteries in service, even on similar cars, will do exactly the same work.

Some cars will cover about 800 km or more each week with long runs, resulting in the battery always being fully charged, and possibly overcharged. Other cars may cover as little as 150 to 300 km each week with short runs, frequent stops and starts, and insufficient running to maintain the battery fully charged. The majority of car owners will possibly cover a weekly mileage somewhere between these two ranges, with batteries maintained generally in a substantially charged condition.

Tests in the laboratory are therefore something of a compromise of the conditions likely to be experienced in service.

The 20-hr capacity test (No. 1) has no great practical significance, although useful in proving the battery as regards the amounts of active materials and volume of acid provided.

The two tests at engine-starting rates of discharge (Nos. 2 and 3) are most valuable in proving that the battery is capable of providing power for starting for a considerable period. A battery which satisfied Test No. 2, at normal temperature, would be capable of supplying at least 70 engine starts each of 5-sec duration taken consecutively without any charge between starts. Similarly, a battery satisfying the test conditions at $-18°$ C (No. 3) would be capable of supplying at least 36 starts each of 5 sec before the battery became discharged. The smaller number of starts at $-18°$ C $(0°$ F) compared with $25°$ C $(77°$ F) demonstrates the loss of capacity which occurs when a battery is cooled. This loss of capacity lasts only while the battery is cold, and capacity is completely restored when the battery temperature is raised to normal. The low-temperature starting test is one of the most exacting of battery tests, and is particularly useful in assessing battery quality.

The test for charge retention (No. 4) when the battery is new proves that the materials used in the battery meet the desirable standards of purity. Impurities in the lead of the grids or paste oxides, or in the sulphuric acid, would increase the loss of charge due to internal chemical action with the battery idle. All lead-acid batteries lose some capacity on standing, and as the loss increases as a battery ages, it is essential that it be kept to a minimum when the battery is new.

A battery in service on a car is alternately discharged and charged. In the winter the demands on the battery are greatly increased, resulting in its stabilizing in a partly discharged condition; in the summer the reverse holds, and the battery tends to be overcharged. The combined effect of winter and summer running conditions is to produce some discharging and overcharging. It is impossible to reproduce exactly these conditions in one laboratory test, and the best compromise is to conduct two life tests.

One of these is the life cycle test (No. 5), where the battery is taken through a fairly deep discharge followed by a generous recharge on each cycle. This test is most useful in proving the quality of the positive and negative active materials (pastes). Batteries not coming up to the desired standard fail this life test owing to shedding of the positive paste or deterioration (sulphation) of the negative paste, or a combination of both.

The second life test (No. 6) is one where the battery is given repeated charges at fairly high temperatures, in an endeavour to reproduce the effects of the prolonged charging to which most batteries are subjected in service. The amount of charge given to the 40-Ah battery on this test would be 600 hr at 4 A, or 2,400 Ah. The test is most useful in proving the corrosion resistance of the positive grid, and batteries which fail show premature collapse of the positive grid frame.

A battery which failed the life cycle test (No. 5) would not necessarily have a short life in service. On the other hand, a battery which failed the overcharge test (No. 6) would almost certainly have a short life. A grid alloy which resists anodic corrosion is therefore one of the most essential features for longevity of automotive batteries.

"Gyp" (cheap) batteries, as supplied mainly for the replacement market, would almost certainly fail most of the above tests. Large manufacturers who supply the initial equipment batteries maintain the same high standards for their replacement batteries.

Discharge Characteristics

The voltage of a 6-cell automotive battery is nominally 12 V, but its precise stabilized voltage on open-circuit is a function of the acid concentration (*see* Fig. 3.3), and varies between about 12·7 V and

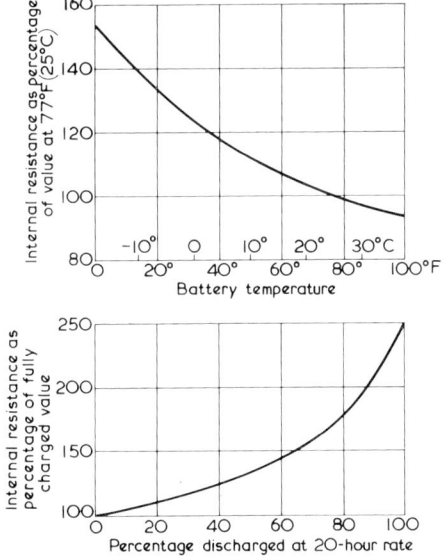

FIG. 6.2. AUTOMOTIVE BATTERIES: VARIATION IN INTERNAL RESISTANCE WITH TEMPERATURE AND STATE OF DISCHARGE

12·5 V. The higher voltage is obtained with batteries having an acid specific gravity of 1·280 and intended for use in temperate climates, and the lower voltage, when 1·250 sp. gr. acid is used for batteries to be operated in ambient temperatures normally in excess of 32° C (90° F).

This voltage of 12·7 V is obtained only so long as the battery is on open-circuit (that is, not delivering or receiving current); when operating on a car the battery voltage will vary between the maximum charge voltage of about 15·5 V and the low value of 8 or 9 V under cold-starting conditions.

The immediate fall in battery terminal voltage on discharge is due to its internal resistance, and is the product of the current (I) and the battery internal resistance (R). Since R is constant at any given temperature and state of discharge, the voltage drop on load is directly proportional to I. The internal resistance of a battery does, however, vary with its state of discharge and temperature, increasing with both depth of discharge and fall in temperature (*see* Fig. 6.2).

VOLTAGE AND CAPACITY VARIATIONS

The decrease in battery voltage with increasing discharge currents is clearly seen in the discharge-voltage/time curves shown in Fig. 6.3; as explained above, this is due to battery resistance. There is also a marked reduction in battery capacity with increasing rate of discharge, and at the 1-min rate the capacity available is only 12 per cent of that obtained at the 20-hr rate. This is due to there being insufficient time for the stronger acid to replace the weak acid in the pores of the plates as the discharge proceeds. Acid diffusion is further restricted by the rapid formation of lead sulphate at the surface of the plates with high discharge currents.

Although 280 ampere-minutes (4·7 Ah), or only 12 per cent of the battery capacity, is available at the 1-min rate, the full rated capacity can always be obtained by reducing the current to the 20-hr discharge value following the high discharge rate. For example, although the battery is exhausted after 1 min at 280 A, and its voltage falls to 8 V or lower, on reducing the current to 2 A the battery voltage immediately rises to about 12 V, and a further discharge for approximately $17\frac{1}{2}$ hr would be obtained at 2 A to give a total output of 40 Ah.

The discharge voltage curves shown in Fig. 6.3 are for the smallest size of battery used on the average size of car, but the same general characteristics would also apply to other reputable makes of battery whatever their capacity.

Capacities at the various rates of discharge can be calculated from the curves and expressed as ampere-hour, watt-hour or percentage capacities as shown in the following table—

Rate	Ah capacity	Percentage Ah capacity	Mean voltage (V)	Wh capacity*	Percentage Wh capacity
20 hr	40	100	11·85	474	100
10 hr	37	92	11·75	435	92
5 hr	32·5	81	11·55	375	79
1 hr	22	55	11·40	251	53
20 min	18·3	47	10·85	199	42
10 min	14·2	36	10·40	148	31
5 min	11·3	28	9·60	108	23
1 min	4·7	12	8·50	40	9

* Ampere-hour capacity multiplied by mean voltage.

A battery of 50-Ah capacity at the 20-hr rate would have capacities at other rates of discharge equal to $\frac{50}{40}$ times those shown in the table. For example, the 20-hr watt-hour capacity would be $\frac{50}{40} \times 474$, or

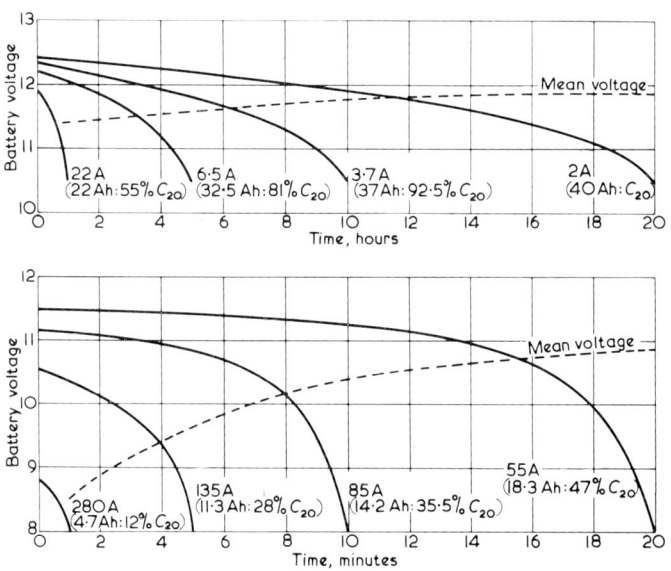

FIG. 6.3. DISCHARGE-VOLTAGE/TIME CHARACTERISTICS OF 12-V 40-AH AUTOMOTIVE BATTERY AT VARIOUS RATES OF DISCHARGE

Battery temperature, 25° C (77° F)

592·5 Wh, and the 1-hr capacities would be $\frac{50}{40} \times 22$ Ah and $\frac{50}{40} \times 251$ Wh. Similarly a 60-Ah battery would have capacities $\frac{60}{40}$ times those shown in the table.

The general shapes of the discharge voltage curves would also be more or less the same for batteries of other capacities.

VOLTAGE/CURRENT CHARACTERISTICS

The curves shown in Fig. 6.4 have been derived from the family of voltage/time curves of Fig. 6.3 by plotting battery voltage after various periods of discharge against discharge current. The plot of

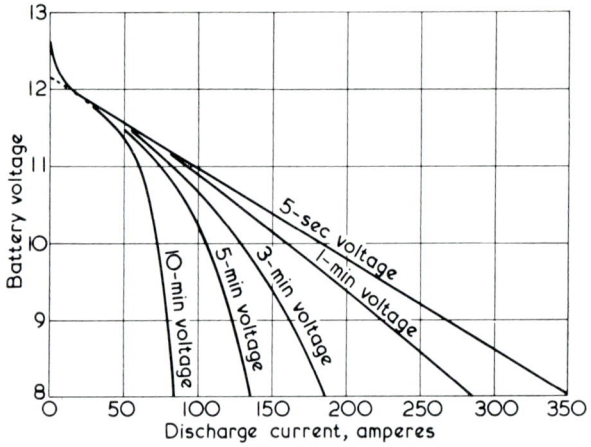

FIG. 6.4. VOLTAGE/CURRENT CHARACTERISTICS OF 12-V 40-AH AUTOMOTIVE BATTERY AT VARIOUS DISCHARGE CURRENTS

Battery temperature, 25° C (77° F). Internal resistance (slope of 5-sec line), 0·012 Ω

the initial battery voltage for the various discharge currents gives a straight line whose slope represents the internal resistance of the 12-V 40-Ah battery. This is found to be 0·012 Ω, which is equivalent to 0·002 Ω per cell. Low internal resistance is most desirable in automotive and other batteries which have to supply large currents for engine starting, as it keeps the voltage drop (*IR*) within the battery low, and ensures the minimum reduction in battery voltage when the load is applied.

Because of the low internal resistance of these batteries, it is possible under certain short-circuit conditions to produce very high currents both inside the battery and through the metal object which short-circuits the battery terminals. If a heavy spanner or tyre lever were jabbed hard across the two terminals of the 12-V 40-Ah battery, it would supply for a second or so a current of the order of 800 A or more. The large power developed by a short-circuited battery is not always realized by motorists, and it is a wise precaution to disconnect the terminal of the battery which is earthed to the car body before commencing any work on the electrical system. Similarly when connecting a battery to the cable terminals, the terminal which is earthed should be connected last of all, as this avoids the unpleasant and surprising cracks and sparks which occur when a spanner accidentally bridges a live contact and the car body. *Contrary to many popular beliefs, the battery plates are not damaged by the passage of heavy currents,* but the heat produced in the intercell connectors by a short-circuit may result in fusing of the lead connector.

Engine Starting

The main concern of most motorists on cold winter days is whether the battery will start the engine, particularly after standing many hours in the open or overnight in a cold garage. Assuming that the engine is in reasonable order and the battery not discharged beyond about 40 per cent of its rated capacity, there should be no trouble in starting. The battery will continue to start the car even in this discharged condition, and at temperatures as low as $-5°$ C ($25°$ F), so long as the running time with the battery on charge is sufficient to replace the drain due to engine-starting and other auxiliary loads. To replace the discharge drain on a battery of one or two engine starts under winter conditions would require a running distance of one and a half miles at a speed of about 50 km/hr. Therefore, unless the battery is discharged beyond the safe limit, or is in poor or worn-out condition, it will be capable of doing its job in the coldest weather without any preliminary help from the starting handle.

Fortunately, most people have sensible ideas about starting handles and recognize that the battery is the motorist's strong right arm for engine starting. The following results of a test taken on a 12-V 40-Ah battery may be of interest to many motorists who think they are extending battery life by occasional use of the handle in place of the battery.

The test was conducted by discharging the battery at 150 A for three 5-sec starts, followed by a charge simulating that of a typical car generator with an input some 15–20 per cent more than the previous output. This was repeated continuously until the battery showed signs of failing, mainly resulting from overheating by the abnormal amount of continuous cycling under car-starting conditions. At this point the battery had supplied 21,000 five-second starts, which would be equivalent to a life of almost 7 years on the assumption that a battery supplies on an average 60 starts per week, or approximately 3,200 per year. It is evident that other factors besides engine starting contribute very largely to the wearing out of a battery in service.

EFFECTS OF TEMPERATURE AND STATE OF DISCHARGE

The curves shown in Fig. 6.5 demonstrate the ability of the automotive battery to provide engine starts for various states of discharge

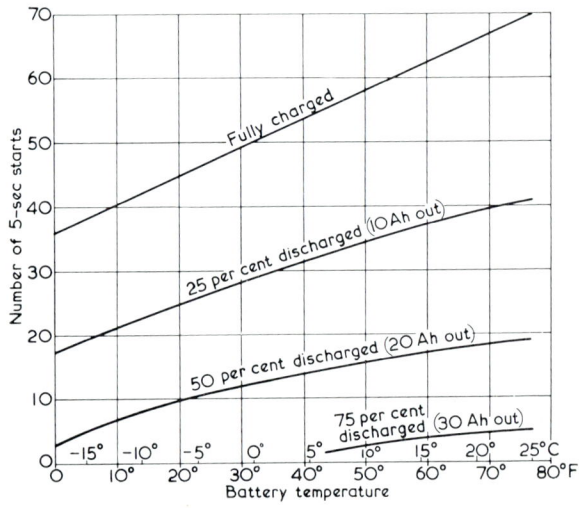

FIG. 6.5. ENGINE-STARTING PERFORMANCE OF 12-V 40-AH AUTOMOTIVE BATTERY IN VARIOUS STATES OF DISCHARGE AND AT DIFFERENT TEMPERATURES

Discharge current, $3 \times C_{20} = 120$ A

and temperature. Although the information is shown for the 12-V 40-Ah battery, it is applicable to any size of automotive battery as the starting current of $3 \times C_{20}$ amperes* is automatically adjusted to battery capacity. It will be seen from the curves that a fully charged battery would provide 70 consecutive 5-sec starts at 25° C (77° F), or 36 consecutive 5-sec starts at −18° C (0° F). As the battery and engine temperatures approached freezing point there would be some increase in the starting current of $3 \times C_{20}$ amperes

* C_{20} = Ampere-hour capacity at 20-hr rate.

owing to the stiffening of the engine. To allow for this it is advisable to fix a safe limit for starting at, say, 10 starts (Fig. 6.5). It is now evident that a battery which had been discharged by about 50 per cent or more of its capacity would not be capable of turning the engine when its temperature fell to about 0° C (32° F). The main conclusions drawn from these curves have been summarized in the following table—

Battery condition	Hydrometer reading (specific gravity)	Engine starting performance in relation to battery temperature
Fully charged	1·280	Excellent even at temperatures well below −18° C (0° F)
25% discharged (75% charged)	1·235	Excellent down to temperatures just below −18° C (0° F)
50% discharged (50% charged)	1·190	Much reduced output. A danger of inadequate power for starting when temperature falls to 0° C (32° F)
75% discharged (25% charged)	1·145	Power available inadequate, even at summer temperature

Determination of Battery Condition

From the foregoing it is obvious that a check of the battery's state of discharge (or charge) should be made particularly before the onset of winter. Batteries which are about 50 per cent discharged or more should be given a charge by a separate charger, as it is highly unlikely that they can be brought to the required state of charge by normal car running. At the same time, it is advisable to check the car's electrical system to determine the reason for the battery's condition. Undercharging can result from a faulty generator, slack driving belts, bad brush contacts, and incorrect regulator settings.

SIMPLE BATTERY-CONDITION TESTS

Three types of instrument are used to check the condition of a battery—the hydrometer, the voltmeter, and the load tester.

HYDROMETER TEST

This test can be made at almost any time, except immediately following the addition of water to the battery, when it would be necessary to give a gassing charge to mix the acid and water before taking readings. The specific gravity of the electrolyte depends on temperature, and is generally referred to a standard temperature of

15·6° C (60° F).   The correction for temperature is 7 points of specific gravity per 10° C variation from 15·6° C, and is added for temperatures above, and subtracted for temperatures below 15·6° C. Unless the battery temperature is extremely low or high, the correction may be ignored for general car battery assessment purposes.

The table below shows the variation in specific gravity with temperature—

Condition of battery	Actual hydrometer readings at temperature of					
	0° F (−18°C)	20° F (−7° C)	40° F (5° C)	60° F (16° C)	80° F (27° C)	100° F (38° C)
Fully charged . .	1·304	1·296	1·288	1·280	1·272	1·264
Half discharged . .	1·214	1·206	1·198	1·190	1·182	1·174
Fully discharged .	1·124	1·116	1·108	1·100	1·092	1·084

VOLTMETER TESTS

To be reliable these tests should not be taken on a battery unless it has been standing idle for at least 12 hr since the last charge (by car generator or otherwise).   The voltmeter should be graduated in hundredths of a volt, for then the voltage readings can be directly related to specific gravity.   For example,

2·12 V per cell approximates to 1·280 sp. gr.

2·08 V per cell approximates to 1·240 sp. gr.

2·04 V per cell approximates to 1·200 sp. gr.

2·00 V per cell approximates to 1·160 sp. gr.

and there is a reduction of about 0·01 V per cell per 10 points fall in specific gravity.   The relationship between open-circuit voltage, specific gravity and battery condition is utilized in the so-called electric hydrometer, which is a special kind of voltmeter with the scale expanded usually between the range of 1·8 and 2·2 V, and calibrated to indicate battery condition.

LOAD TEST

Load testers usually consist of two prongs bridged by a resistance bar across which is connected a voltmeter with a coloured scale divided into green, yellow and red zones.   When making the test the prod points are pressed hard across the terminals of each cell for a few seconds.   The current flowing in the resistance bar can vary from tens of amperes to a hundred or more amperes, depending on the design of instrument, but the common intention is to read and note the voltage of each cell under load, and compare the readings.

In the event of a battery being tested immediately off charge, the surface charge should be removed before load testing. This can be done by leaving the headlights or similar electrical load switched on for 2–3 min, after which time the test can be made.

DIAGNOSIS OF BATTERY CONDITION FROM HYDROMETER, VOLTAGE AND LOAD TESTS

Uniformity of readings of individual cells is usually a sign of a healthy battery, even though the level of readings may indicate that the battery is discharged.

A difference in individual cell readings of more than 0·05 V or 50 points of specific gravity is a sign that the battery is reaching the end of its useful life.

Diagnosis Chart of Battery Condition

Observed readings			Battery condition	Recommendations
Hydrometer test	Voltage test	Load test		
Readings uniform and within the range 1·240–1·270	Readings uniform and within the range 2·08–2·11 V	Meter needle indicating steady value in "green" zone	Healthy and in a reasonably charged condition	No action required
Readings uniform and within the range 1·180–1·200	Readings uniform and within the range 2·02–2·04 V	Meter needle indicating steady value in "yellow" zone	Healthy but half discharged	Suggest a bench charge
One cell about 30 points or more lower than remainder	One cell about 0·03 V lower than remainder	Meter needle falling during test of one cell	May be (a) cell out of step, or (b) faulty cell	(a) May be rectified by bench charge
Readings irregular and 50 points or more variation	Readings irregular and 0·05 V or more variation	Meter needle falling during test and in "red" zone	Battery at end of useful life	Replace

Any one of the three tests shown in the table would provide the information required for assessment of battery condition.

Charging on the Car

For an automotive battery to remain healthy and give maximum life in car or motor vehicle service, it must operate most of the time in or near the fully charged condition. The engine-driven generator which supplies direct current for charging the battery and other loads must therefore have an output at least equal to the total load, so that there is some current available most of the time for charging the battery.

When the battery is discharged by about 50 per cent or more of its capacity, there is a danger that it cannot be recharged to the desired state by charging under normal running conditions. This situation is, of course, aggravated during the winter months, as the balance of current available from the generator for battery charging is reduced owing to the increased electrical load. Under such conditions the battery should receive a supplementary charge from a separate source such as a "home" charger, when the car is in the garage. This does not mean that every battery requires "bench" or "home" charging, as even in the winter, with a correctly balanced electrical system, the input charge should almost equal the output, providing the running time during which the battery is charging is adequate. This additional charging should therefore be necessary only where abuse of the battery in some form or other has occurred, or where faults may have developed in the generator or electrical system.

The charging current available from the generator will depend on the load on the generator, and the difference between full output current and load current will be available for charging the battery. A discharged battery will accept the full output of the generator until the battery voltage begins to rise, when the current falls gradually to 1–2 A, and finally decreases to a fraction of an ampere when the battery is fully charged. This automatic regulation of the generator output to suit the battery condition is the principle behind the compensated voltage-control system.

Compensated Voltage-Control Generator System

The compensated voltage-control equipment, consisting of a specially designed generator, regulator and cut-out unit, has been used on cars and light vehicles since about 1937. Previous to that date the third-brush generator system had been used, but, because of its virtually constant-current output, it had the serious disadvantage of overcharging batteries at high currents, with the inevitable reduction in battery life.

Fig. 6.6 is a schematic diagram of a typical compensated voltage control unit. The regulator and cut-out, although housed in the same box, are separate units. The cut-out has a shunt and series winding, and operates by closing the contacts between generator and battery circuit when the generator voltage builds up to about 13–13·5 V. The series coil assists in keeping the contacts closed by current in the forward direction. If the generator voltage falls below the battery voltage a reverse current flows which weakens the shunt field, causing the contacts to open.

The regulator windings consist of a voltage shunt winding across the generator and two series current windings. One of these carries the generator full output current and the other carries the ignition, lighting and accessory loads.

CONSTANT VOLTAGE CONTROL

In operation, when the generator voltage rises to a predetermined value the voltage shunt winding is energized sufficiently to attract the regulator armature. This results in the regulator contacts opening

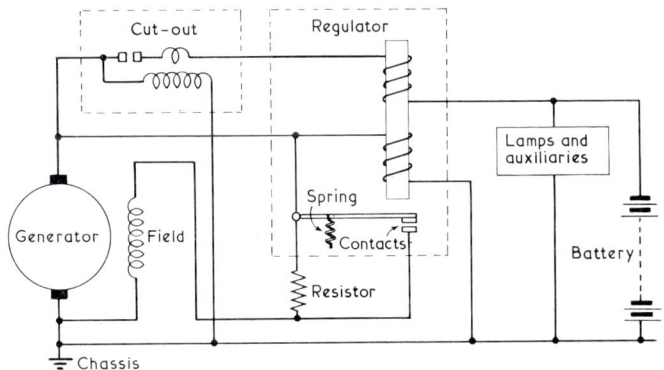

FIG. 6.6. GENERATOR SYSTEM WITH COMPENSATED VOLTAGE CONTROL

and inserting the resistance in the field circuit, so that the generator field current is reduced. The generator voltage falls, the regulator shunt-winding field is weakened, and the regulator contacts close again, causing the generator voltage to build up again to maximum, and the whole cycle is repeated continuously with the contact arm in a state of rapid vibration. The amplitude of these vibrations is increased with increasing generator speed, and the average value of the field current is reduced, with the result that the output voltage of the generator is practically constant above a speed of about 40 km/hr.

LOAD COMPENSATION

In the event of the combination of a heavy load and a discharged battery, the regulator series winding carrying the load current would work along with the voltage coil and reduce the operating voltage of the regulator. This would protect the generator against overload by limiting the output current.

TEMPERATURE COMPENSATION

To compensate for temperature differences which produce voltage changes in the battery being charged—a cold battery has a higher on-charge voltage, and a hot battery has a lower one compared with a battery at $21°$ C ($70°$ F)—a bimetallic strip is incorporated in the regulator armature. This reduces the regulator voltage when hot and increases it when cold, so that the generator voltage follows to some extent the corresponding changes in battery voltage.

VOLTAGE SETTING OF REGULATOR

Most regulators for use in temperate climates are set by the manufacturer about 15–15·5 V, which represents the average voltage on charge of a battery in a fully charged condition at normal temperature. This voltage setting is increased for cold climates and decreased for warm climates to suit the corresponding changes in battery voltage as mentioned previously.

The great diversity of operation of cars could mean that the voltage setting as made initially does not necessarily match the particular operating conditions. For instance, a car which is covering big mileages requires a lower regular setting to prevent overcharging of the battery.

A sure guide to the correct setting of the regulator for any particular operation is the amount of water consumed by the battery. The optimum regulator voltage setting is that which maintains the battery near to the fully charged condition with the minimum amount of water. In this country an automotive battery should require topping up with water only about every 1600 km. This water replaces that lost mainly owing to gassing, when the water in the electrolyte is converted to hydrogen and oxygen; 100 Ah of gassing charge accounts for 34 cm³ of water. In addition, there would be some slight loss of water due to evaporation, but this is negligible compared to that lost by gassing the battery on charge.

Regulator voltage settings should be checked periodically and adjusted where necessary by a competent automobile electrician. Properly adjusted regulators ensure correct charging in service and long, trouble-free battery life.

BATTERY AND GENERATOR CHARACTERISTICS DURING CHARGING

A typical set of curves showing the generator voltage and current during the recharging of a battery is shown in Fig. 6.7. The time T for the battery to reach its stabilized, fully charged voltage will depend on the initial state of the battery. The shapes of the curves will be generally similar, but the time T will be short (minutes) when the battery is initially almost fully charged, and long (hours) when the battery is 50 per cent discharged or more.

It will be appreciated that the curves are shown for a battery load only; their shape would be affected somewhat by the presence of other loads, owing to the reduction in available charging current.

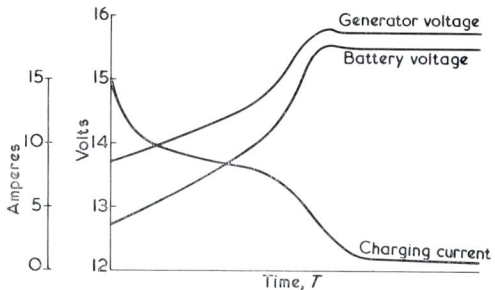

FIG. 6.7. TYPICAL CURRENT/TIME CHARACTERISTICS OF GENERATOR WITH COMPENSATED VOLTAGE CONTROL WHEN CHARGING 12-V 40-AH BATTERY

Generator speed, 2,000–2,500 rev/min

Automotive Batteries for Heavy Vehicles

The various classes of vehicle under this heading comprise medium and heavy commercial vehicles, coaches, passenger service vehicles, tractors, etc. All require a battery for engine starting, lighting and other duties, and because the demands on the battery are heavy owing to the use of Diesel engines and considerable parking loads, the battery has to be more robust, of greater capacity and longer life expectancy than batteries used for private cars and light commercial vehicles.

HEAVY-DUTY BATTERY SYSTEMS

The starting of Diesel engines presents a greater problem than that of petrol engines. High compression and high engine turning speeds are necessary to ignite the fuel, and at low temperatures it is desirable to pre-heat the fuel and air passing to the cylinders. All three conditions impose a considerable drain on the battery. On starting, the heavier Diesel engines may need a current of 600–800 A for breakaway, and a spinning current of 300–400 A at 16 V for up to 20–30 sec. To provide this current with the battery in a 25 per cent discharged condition, and at temperatures down to $0°$ C ($32°$ F), requires a battery of a hundred or more ampere-hours at a nominal 24 V.

BATTERY SIZES AND DESIGN

Batteries vary considerably in capacity, size and design. The capacity at the 20-hr rate varies from about 80 Ah to 350 Ah, the smallest sizes being assembled as 6-cell 12-V units, and the larger sizes, because of their size and weight, as 3-cell 6-V units. All the heavier vehicles use a 24-V battery system.

Battery design can be divided into three main classes—

1. Batteries of minimum weight and compactness for use on commercial vehicles and the smaller types of passenger vehicle.

2. Heavy robust batteries designed for maximum life.

3. Batteries of design intermediate between Classes 1 and 2.

FIG. 6.8. HIGH-PERFORMANCE BATTERY FOR PUBLIC SERVICE VEHICLES

Class 1 batteries use comparatively thin flat pasted plates approximately 2·5 mm thick in conjunction with the dual separation of a glass-wool mat and porous rubber or plastic separator. Capacities range from about 80 to 150 Ah.

Class 2 batteries have flat plates about 6 mm thick and dual separation as in Class 1. Alternatively, the positive plates may be of the armoured tubular type similar to those used in traction cells, a single separator being used with a negative plate approximately 5 mm thick. Capacities range from 120 to 350 Ah.

Class 3 batteries have flat pasted plates about 3·5 mm thick with dual separation (Fig. 6.8). They are designed to give increased starting performance for less weight and volume compared to the batteries in Class 2. The range of capacity is 150 to 270 Ah.

PERFORMANCE

Heavy-duty automotive batteries with thicker plates and dual separation have a somewhat higher internal resistance per ampere-hour than car batteries. When the ampere-hour output at the various rates of discharge is expressed as a percentage of the capacity at the 20-hr rate, there is a steady reduction at the quickest rates compared with that for the car battery. This is shown in the following table for a battery temperature of 25° C (77° F).

Ampere-hour Capacities at Various Rates as Percentages of 20-hr Capacity

| Rate | Car battery | Heavy-duty battery | | | Final voltage per cell |
		Class 1	Class 2	Class 3	
	per cent	per cent	per cent	per cent	V
20 hr	100	100	100	100	1·75
10 hr	92	92	92	92	1·75
5 hr	81	81	80	81	1·75
1 hr	55	54	49	52	1·75
30 min	47	44	39	42	1·33
10 min	36	32	27	30	1·33
5 min	28	26	20	23	1·33
1 min	12	9	5	7	1·33
1-min amperes per 100 Ah	720	540	300	420	1·33

It will be appreciated that, for the same 1-min capacity, a heavy-duty battery of Class 2 would need to have a 20-hr capacity 2·4 times the 20-hr capacity of the car battery.

Life

Battery costs are very important to the large fleet operator, and frequently batteries which are the most expensive on initial outlay (Class 2) prove to be the most economic in terms of cost per annum. It is, however, difficult to forecast battery life, as so much depends on mileage as well as maintenance and attention to correct charging techniques in service. Estimated average lives for reputable batteries under Classes 1, 2 and 3 would be 4, 8 and 6 years respectively.

Service Demands

The electrical accessories have steadily increased in the heavy vehicle field so that the battery has to provide more current for longer periods. To maintain the battery in the optimum state of charge would mean long hours of running with the compensated-voltage control system as used on car and light vehicles, with the charging current decreasing steadily from the time of commencement of charge. The time taken to recharge the battery can be reduced by maintaining the output current of the generator substantially constant until the battery reaches its gassing voltage, when the current falls off to a safe low value. This method of charging is achieved by the current-voltage controlled generator.

Charging Systems

D.C. Generator

With current-voltage control there is regulation of both current and voltage by means of two regulator units. One unit (current

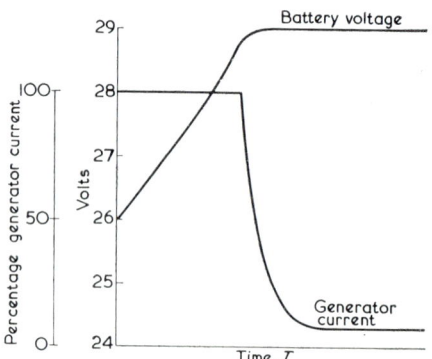

FIG. 6.9. TYPICAL CHARGE CHARACTERISTIC WITH CURRENT-VOLTAGE CONTROLLED GENERATOR

control) keeps the generator current constant until the battery voltage reaches a certain value. At this point the second regulator unit (voltage control) maintains constant generator voltage and the charging current falls with rise in battery voltage. This charge characteristic is shown in Fig. 6.9, where the voltage regulator takes control at a battery voltage of 29 V for a 12-cell battery.

A.C./D.C. System

In congested towns and cities, passenger service vehicles run for considerable periods with the engine at idling speed and the generator out of circuit. In these circumstances the batteries will work in a discharged condition, and usually the balance of charge can only be restored by giving periodic bench charges.

This trouble has been overcome largely by the use of alternator/rectifier (a.c./d.c.) systems. By suitable choice of drive ratio the a.c.

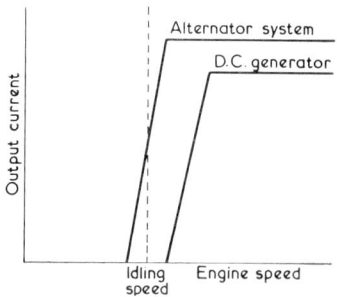

FIG. 6.10. COMPARISON OF TYPICAL A.C./D.C. AND D.C. GENERATOR
OUTPUTS

generator provides a current at engine idling speeds and can tolerate the considerable increase in speed at maximum engine revolutions. This is not possible with the d.c. generator, as the drive ratio necessary to provide a current at idling speeds would mean that at maximum engine revolutions the d.c. generator was running well above its maximum permissible speed. The maximum speed of the d.c. generator is limited by commutation problems, high temperatures and arcing at the brushes. These problems do not exist with the a.c. generator.

A comparison of d.c. and a.c./d.c. systems is shown in Fig. 6.10, where it can be seen that over half the output of the a.c./d.c. system is available at idling speed.

The use of a.c. generator/rectifier systems is likely to extend to other vehicles including police cars, taxis, ambulances, etc., where the installation of radio equipment imposes a standing load on the battery which cannot be replaced by the d.c. generator system.

The vibrating-contact type of regulator as used with d.c. generators can be used to regulate the output of the alternator by field control. The regulation is by current and voltage similar to that shown for the d.c. generator in Fig. 6.9.

Transistorized regulators are also being used, and with the elimination of contacts and moving parts, have greater reliability and require less maintenance than the vibrating type of regulator.

Charging off the Vehicle

BENCH CHARGING

Any charge given to the battery off the car or vehicle is referred to as a *bench charge*. This is usually necessary when the battery has been abused in some way such as prolonged parking with the lights on, faults in the electrical system, or inadequate charging in service. To be suitable for bench charging, a charger should have an output of several amperes, and usually a rating in amperes equal to about 5 per cent of the battery capacity is suitable for most batteries.

A charge for a single night, or about 15 hr, should be sufficient to restore the battery to a substantially charged condition. For a charge to be correctly administered the following instructions should be observed.

INSTRUCTIONS FOR CHARGING

1. Check the acid level in each cell and top up with pure water when the levels are below the separators.

2. Replace the vent plugs, and ensure that the vent holes are free to permit the escape of gases evolved during the charge.

3. Connect the positive lead of the charger to the positive terminal of the battery (marked "+" or "P"), and the negative lead to the negative terminal of the battery.

4. Switch on the charger, and check that the charging current is passing by observing the ammeter reading.

5. Continue the charge until all cells are gassing freely and voltage and specific gravity readings have reached maximum values over a period of 4 or 5 hr.

6. The specific gravity readings should be within 10 points of the recommended fully charged value (usually 1·280 at 15·6° C (60° F)).

7. Switch off the charger, and allow the battery to stand for several hours.

8. Check the acid levels. If they reach only to the tops of separators or are lower, add more water and give a further charge of one or two hours to mix the water and acid.

9. Disconnect the battery from the charger, and clean the top of the battery with a wet cloth to remove any acid or dirt. Finish off by wiping with a clean dry cloth.

FAST CHARGING

Fast chargers have become popular because they enable garages to give a quick emergency recharge. Most of these chargers provide an output of about 50 A, tapering to 40 A, during the recharging of 12 V batteries, and about double this output for the larger-capacity 6-V batteries. A healthy discharged battery can be brought to within 80–90 per cent of full charge in about 45 min when charged by a fast charger.

Because heavy gassing and a sharp rise in battery temperature occur on each fast charge, it is advisable to limit the number given to a battery to about a dozen during its life. If more than this are required there is something wrong with the method of operation which should be traced and corrected.

An unhealthy battery can be damaged by a high charging current. If the battery is known to be in poor condition or sulphated, or has been left in a discharged condition for a considerable time, it can only be restored by a long slow recharge. The condition of the cells must therefore be checked beforehand to see whether the battery is suitable for fast charging.

It is highly desirable for some automatic means of limiting the duration of a fast charge to be incorporated in the charger. This may take the form of a thermal switch or thermostat inserted in one of the centre cells of the battery, and set to cut off the charger when the battery temperature reaches 46°–51° C (115°–125° F).

An alternative method is to provide the charger with a time switch which terminates the charge after a suitable period. This period is determined by the original state of discharge of the battery as obtained by hydrometer readings or other suitable tests.

Maintenance

ACID LEVELS

Check the acid levels periodically, about once a fortnight or every 800 km, and add pure water as necessary. Never add acid.

Care should be taken not to exceed the normal level, which is usually about 6 mm above the tops of the separators. If a battery is overfilled there is a danger of the acid overflowing through the vent plugs when the battery is charged. Acid from the battery will attack hold-downs and other metal parts in its vicinity.

Most mains water is sufficiently pure for topping up automotive batteries, except in certain areas where the water is chlorinated or of high mineral content.

CORROSION

Corrosion is due to the action of acid on metallic parts such as terminals and hold-downs. It can be prevented by keeping the top of the battery clean and dry, and protecting metal surfaces likely to be affected by a coating of petroleum jelly. Whenever acid has been spilt the affected areas should be immediately wiped with a rag soaked in ammonia. If corrosion has already occurred the areas should be cleaned by scraping, washing and drying, and protected by applying a coating of petroleum jelly.

In cases of poor starting, check for terminal corrosion and tightness of terminal connexions, as high-resistance connexions in the battery circuit often stem from corroded or loose-fitting terminals.

LAYING UP BATTERIES

When not in use, a battery loses some of its charge by internal chemical action, and this loss of charge is increased by high temperatures or external leakage currents if the battery top is wet and dirty. Batteries which remain long in a low state of charge become sulphated beyond the stage where they can be restored by recharging. The following procedure should be followed to maintain batteries healthy during laying-up periods—

1. Disconnect the battery from the electrical system of the vehicle.
2. Check the acid levels, and add pure water as necessary to restore levels which are low.
3. Give the battery a full recharge as under "Bench Charging" (page 130).
4. Clean and dry the battery top and terminal posts.
5. Store in a cool, dry place.
6. Recharge fully at least every two months and immediately before putting into service.

SAFETY

Never bring a flame, lighted cigarette or pipe near the battery during or shortly after a charge, as the gases generated on charge are a mixture of hydrogen and oxygen. The danger of an explosion is very remote under normal circumstances, and the risk is reduced if the cell vent plugs are tightly in position.

Care should be taken to see that the vent holes do not become blocked with dirt or grease.

CHAPTER 7

STATIONARY BATTERIES

STATIONARY batteries as a class include those using either Planté or pasted positive plates. This chapter deals solely with Planté batteries, as it is this type which is used extensively for most stationary battery applications in this country.

The history, design and characteristics of the Planté cell have been described previously, but it is worth while to recall the considerable number of different sizes and types used to suit a wide variety of duties.

Plate sizes range from 7·5 Ah (7·6 cm × 12 cm) to 432 Ah (51 cm × 76 cm), with dozens of intermediate sizes. Plate thickness varies between 8 mm ($\frac{5}{16}$ in.) to almost 12 mm ($\frac{1}{2}$ in.) for positive plates, and between 5 mm ($\frac{3}{16}$ in.) to 8 mm ($\frac{5}{16}$ in.) for negative plates.

Planté cells are of two main types: open type (Fig. 7.1) and enclosed type (Fig. 7.2).

Open-type cells (cells without top covers) are available in all sizes and are assembled in glass or lead-lined wood boxes. Glass boxes are available up to 1,000 Ah, and lead-lined wood boxes from 400 Ah upwards.

One of the largest cells supplied is just over 15,000 Ah, weighs 4000 kg, and measures 132 cm (h) × 170 cm (l) × 69 cm (w).

All open-type cells have to be assembled, filled and charged on site.

Enclosed-type cells (those which are fitted with a hard-rubber, plastic or composition cover) are available in sizes from 10 Ah to 2,000 Ah in glass boxes, and the range is likely to be extended to meet the demand. Up to the 400-Ah size, these cells are supplied to the customer in a filled and charged condition.

The enclosed-type cell has superseded the open type in almost all the smaller and intermediate applications.

INSTALLATION

Small-capacity batteries of the enclosed type may be installed in a single-unit steel cabinet together with the charger and control unit (Fig. 7.3). For larger batteries it is usual to mount the cells on wood stands or stillages at a convenient height above floor level (Fig. 7.4).

Open-cell batteries are usually located in a separate battery room

Fig. 7.2.   High-performance Planté Cells in Glass Boxes

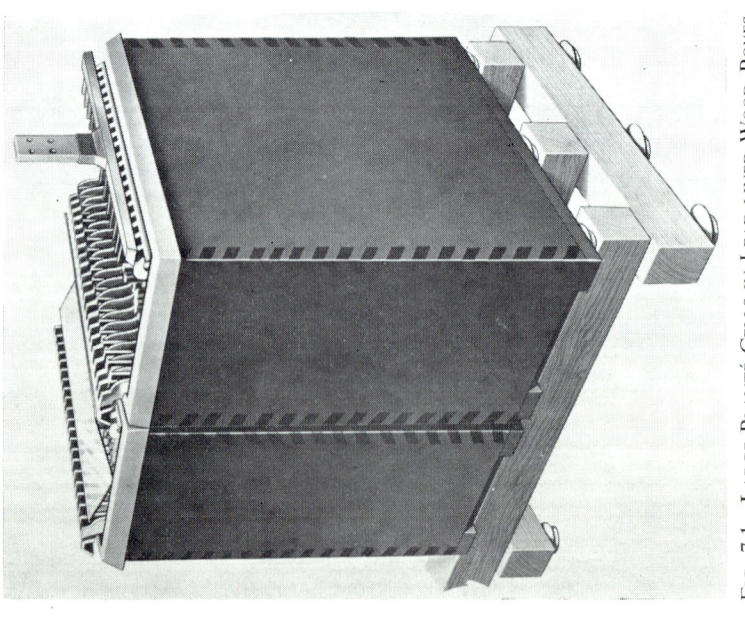

Fig. 7.1.   Large Planté Cells in Lead-lined Wood Boxes, Mounted on Stillage

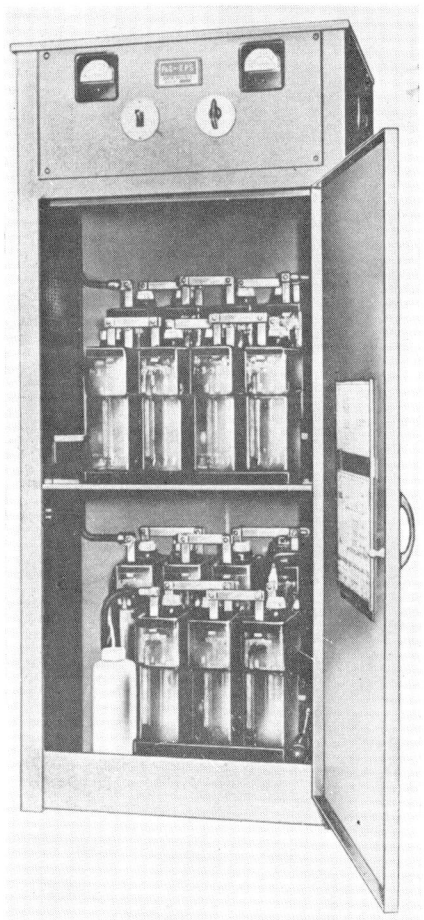

FIG. 7.3. BATTERY OF PLANTÉ CELLS HOUSED IN STEEL CUBICLE, FOR SWITCH-TRIPPING DUTIES

apart from other electrical equipment. Enclosed-cell batteries may be operated in close proximity to other gear without any danger.

Battery Rooms

The battery room should be well ventilated, dry and as moderate in temperature as climatic conditions permit, with all cells operating at about the same temperature. Cells should therefore be protected

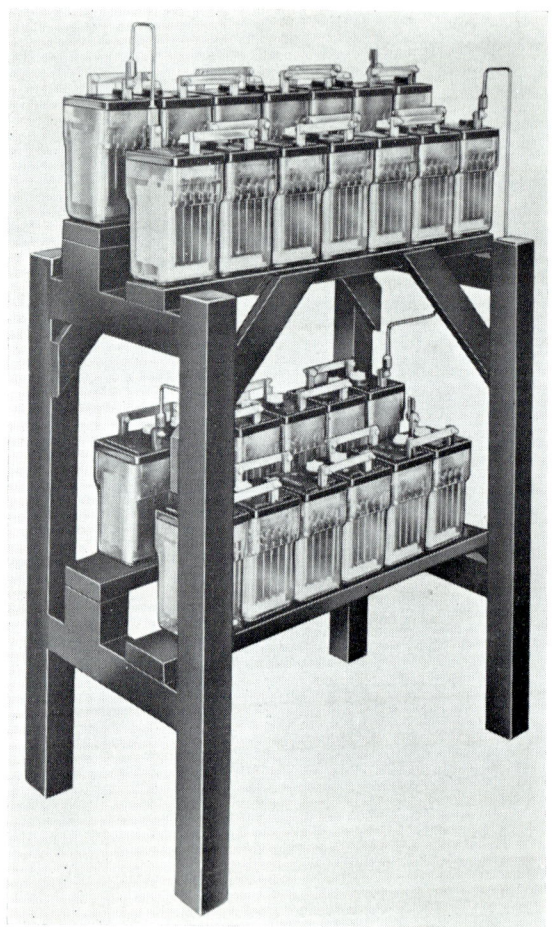

FIG. 7.4. PLANTÉ BATTERY, MOUNTED ON WOOD STAND,
WITH DOUBLE-ROW DOUBLE-TIER ASSEMBLY

from direct sunlight, by screening or white-washing windows where necessary.

Lighting. The battery room should be well lighted to facilitate maintenance and inspection of the battery.

Ventilation. Good ventilation is important to keep the room dry and to avoid the concentration of gases given off during charging. If fans are used, it is advisable to mount them high on an outside wall, so that they expel to the outer air and take in air from inlet vents set low on an opposite wall.

Floors. Batteries of the enclosed type can be installed in rooms with dust-free concrete floors. For the larger open-cell batteries, it is advisable to use an acid-proof floor surface such as asphalt.

Supporting the Battery

There are three recognized methods of supporting stationary batteries: stands, stillage, and pedestals.

STANDS

Stands are suitable for glass-box batteries, and may be single or double tier. For the larger wood-box cells, single tiers and single rows are normally used.

It is important that the stands should be made so that each cell is accessible for adding water or taking readings when required. The cells should be evenly and firmly seated on the stands, which are stiffened when necessary by bracing struts between the stand legs and runners carrying the cells (Fig. 7.4).

STILLAGE

This form of support is made by laying a series of wooden cross-bearers on the floor or on insulators, and then fixing long runners, which will carry the cells, at right angles to the cross-members. Stillage support is used mostly for the large open-top cells (Fig. 7.1). All timber which is used for battery stands is given several protective coatings of anti-sulphuric paint, or enamel.

INSULATORS

For batteries of over 75 cells mounted on stands, insulators are fitted under each cell. No insulators are used for up to and including 60 cells, but for over 60 cells, insulators are fitted under each stand leg. These are general rules only, and insulators may be fitted for any battery arrangement at the discretion of the customer or battery manufacturer.

PEDESTALS

These are mounted directly on the floor or on tiles placed on the floor. The cell box is placed direct on to the pedestals and is carefully levelled.

BATTERY LAYOUT

The placing of stands, stillage or pedestals depends on the size and shape of the battery room, but all arrangements should allow easy access to every cell in the battery. A minimum width of gangway of 2 ft should be allowed to enable cells to be withdrawn should it be necessary to carry out a repair.

Method of Operation

The Planté battery will work efficiently and give a long life, over a wide range of operating methods and conditions.

A list of typical operating methods is given below—

(*a*) *Straight Charge/Discharge*. Where the charging source is not habitually "on" when the battery is discharging. The load *may* be on during charge.

(*b*) *Assisted Charge/Discharge*. Where it is usual to run the charging source whenever the load is heavy.

(*c*) *Straight Trickle Charging*. Where the only discharge from the battery is either a failure of supply or a test discharge.

(*d*) *Compensating Trickle Charging*. Where the trickle-charge current is higher than the normal to make up for periodical discharge such as switch operation, etc.

(*e*) *Floating and Trickle Charging*. Where the battery is normally working on the floating system but is left on trickle charge when not in use.

(*f*) *"Equilibrium" Floating*. Where the cell voltage is 2·05–2·10 V and the specific gravity readings therefore fall day by day. Freshening charges are required weekly or more often.

(*g*) *Straight Floating*. Where the cell voltage is 2·10–2·20 V, so that discharge is prevented and a partial compensation is provided for normal open-circuit losses. Freshening charges are required occasionally.

(*h*) *Floating/Trickle Charge*. Where the cell voltage is 2·20–2·30 V (in some cases even higher), so that the discharge is prevented and full compensation is given for local-action losses. If maintained continuously, especially if by automatic charger of "Cycloc" type, freshening charges are very rarely required.

Regular charge/discharge operation is now almost obsolete, but was used extensively for telephone and house lighting installations up to about 1940.

Applications

There are three main applications for stationary batteries, namely: power stations and substations, telephones, and emergency lighting.

Power Stations

In the early days of electric power stations, very large storage batteries were installed as an essential part of the power plant. Then, all electrical power generated and distributed was direct current, which enabled batteries to be connected directly to the distribution system. By this means, batteries assisted in fuel and running economies and maintenance of supply, and were subjected to fairly regular cycles of discharge and charge.

As the electrical industry rapidly developed at the start of the twentieth century, the advantages of high-voltage a.c. supply and distribution meant that many of the original d.c. stations were gradually replaced. This change from d.c. to a.c. supply introduced many new problems associated with the high voltages used, as all equipment had to be made "safe" from the point of view of electric shock.

Remote control of switches and equipment was introduced, and because of their reliability and flexibility, batteries were considered to be the best source of supply for operating circuit-breakers and the many safety and protective devices which are part of the complexity of a.c. generation and distribution through the national grid system.

Modern power stations and substations require a number of batteries of different sizes to operate a variety of duties. The nuclear power stations require larger batteries than the conventional stations for essential duties in order to maintain the maximum degree of safety at all times.

Typical examples of battery applications in modern power stations of both types, are given below.

CONVENTIONAL POWER STATION BATTERY INSTALLATION: HIGH MARNHAM POWER STATION (NEWARK, NOTTS)

(a) 240-*volt batteries*

Two batteries each 113 cells of 2,000 Ah at 10-hr rate (904 kWh).
Duty. In an emergency, batteries to be connected in parallel to supply standby motor auxiliaries, emergency lighting, and switch operation. Battery to be capable of giving 859 A for 4 hr to a minimum voltage of 192 V (1·7 V per cell).

(b) 110-*volt batteries*

Two batteries each of 55 cells of 400 Ah at 10-hr rate (88 kWh).
Duty. Each battery to supply indication and control circuits. Battery to be capable of giving 43 A for 6 hr to a minimum voltage of 104·5 V (1·9 V per cell).

(*c*) 50-*volt batteries*

Two batteries each of 24 cells of 200 Ah at 10-hr rate (19 kWh). *Duty.* In an emergency, batteries to be connected in parallel to supply telephones, indicating and alarm equipment. Battery to be capable of giving 41 A for 6 hr to a minimum voltage of 46·3 V (1·93 V per cell).

(*d*) 50-*volt battery*

One battery of 24 cells of 60 Ah at 10-hr rate (3 kWh). *Duty.* Coal plant alarms and indicators.

All the batteries under (*a*) (*b*) and (*c*) consist of Chloride Planté open-type cells constructed to British Standard No. 440: 1932. An illustration of part of a typical battery installation for power stations using these cells is shown in Fig. 7.5.

NUCLEAR POWER STATION BATTERY INSTALLATION: SIZEWELL (SUFFOLK)

(*a*) 440-*volt battery*

Two batteries in parallel each of 224 cells, 1,300 Ah at 10-hr rate (1,144 kWh). *Duty.* All essential emergency supplies for reactor. Battery to be capable of giving 906 kW for 15 min to a minimum voltage of 390 V (1·75 V per cell).

(*b*) 240-*volt battery*

One battery of 120 cells, 210 Ah at 10-hr rate (50 kWh). *Duty.* Emergency oil seal and fluid oil pumps, lighting and indication. Battery to be capable of giving 70 kW for 15 min to a minimum voltage of 200 V (1·67 V per cell).

(*c*) 110-*volt battery*

One battery of 55 cells, 1,200 Ah at 10-hr rate (132 kWh). *Duty.* Switch operation and indication. Battery to be capable of giving 100 A for 10 hr with 253-A load superimposed. Minimum voltage, 96 V (1·75 V per cell).

(*d*) 50-*volt battery*

One battery of 24 cells, 200 Ah at 10-hr rate (10 kWh). *Duty.* Automatic telephone exchange and station alarms.

All the above batteries consist of Chloride high-performance Planté enclosed-type cells.

Fig. 7.5. Section of Large Planté Battery for Power Station Duties

Comparing the total battery capacity required for the two stations, it will be seen that the nuclear station uses 30 per cent more battery power than the conventional power station. This difference would have been very much greater had the heavy type of Planté cell been used for the nuclear station as was used for the older, conventional station. The advantage of the high-performance Planté cell is shown in the discharge curves at the 10-hr to 1-hr rates in Fig. 7.6. At rates

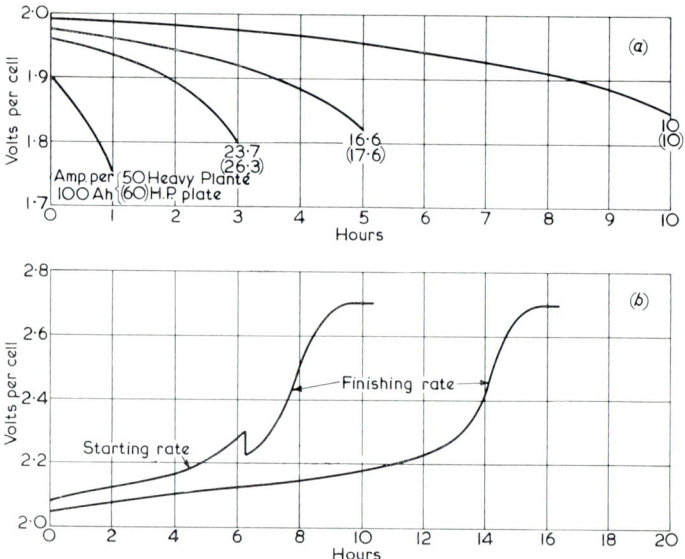

FIG. 7.6. TYPICAL DISCHARGE AND RECHARGE VOLTAGE/TIME CHARAC-
TERISTICS OF PLANTÉ CELLS (HEAVY AND HIGH-PERFORMANCE TYPES)

(a) Discharge characteristics at constant current
(b) Recharge characteristics following discharge at 10-hr rate

quicker than the 1-hr the performance of the high-performance cell is very much greater than that of the heavy-plate cell. For example, duty (a) (Sizewell power station) can be met by a 2,600-Ah high-performance Planté battery. The battery capacity would have to be increased by about 50 per cent, that is, in the region of 4,000 Ah, if a heavy Planté battery were used.

SUBSTATION BATTERIES

The large attended substations have batteries installed to supply similar standby duties as required by the conventional power stations. Small unattended substations usually require a relatively small battery for switch-tripping duties.

Telephone Systems

A public telephone system relies on a continuous power supply in order to provide a 24-hr service. Power is provided by the public electricity supply, and when that fails, interruptions are avoided by reserve power supplied by batteries and engine-generators.

The British Post Office spends a considerable amount of money on Planté batteries for telephone service, and is in fact the largest single purchaser of this type of stationary battery. These batteries are an essential part of a telephone system as they ensure a continuous power supply at all times. An automatic telephone exchange cannot tolerate even a momentary break in supply, because this would produce disastrous effects on the dial impulse circuits and other equipment. The surge of current, which would be very considerable, when the supply was restored would also blow the equipment fuses, and prevent automatic restoration of service.

Large storage batteries were first installed in telephone exchanges about 1900, when the common or central battery (C.B.) system was introduced. The large central battery of lead-acid cells supplied power for all the telephones connected to the one exchange, and replaced the older system in which each subscriber's instrument had a separate primary battery.

CHARGE/DISCHARGE SYSTEMS

Up to the 1930s, most telephone exchange batteries operated on a charge/discharge routine. The capacity available from the battery had to be at least equal to the daily load, and because of regular deep cycles, battery life was limited to about 6–8 years.

By about 1938, charge/discharge systems were superseded in most exchanges by a floating battery arrangement. Some of the early schemes included counter-e.m.f. cells which were switched automatically in the load circuit when the battery was on charge. The counter-e.m.f. cells reduced the high voltage of the battery on charge to below the maximum permissible value of 52 V. (A description of counter-e.m.f. cells will be found on page 224.)

FLOATING-BATTERY SCHEMES

The larger telephone exchanges now standardize on a 25-cell floating battery because of its good discharge-voltage characteristic and general reliability of the system. The standby capacity is provided in two batteries of equal size, connected in parallel across the exchange load and rectifier or generator supply. Each battery is disconnected in turn from the circuit and recharged at fortnightly intervals.

The characteristics of the 25-cell battery on telephone floating duties, and the considerations which make it necessary for each battery to be disconnected in turn and charged from a separate supply, are worth noting. For 50-V telephone exchanges the equipment limits are 46–52 V. For a 25-cell battery these figures represent 1·84 and 2·08 V per cell respectively. The upper limit is only slightly higher than the open-circuit value of 2·05 V for a Planté cell, and when floated at this voltage the battery slowly discharges. During fluctuations of peak loads the battery also provides some power to the telephone system, so that over a period it becomes partly discharged.

To restore the battery to the fully charged condition the charge current towards the completion of charge must be of a value which will elevate the battery voltage to the equivalent of at least 2·6 V per cell. This is equal to at least 65 V for the 25-cell battery, which is far in excess of the maximum limit of 52 V for the exchange equipment.

If 24-cell batteries are used the limits are 1·92–2·17 V per cell. On discharge, the battery voltage falls to the equivalent of 1·92 V per cell before the full available capacity can be used, and larger batteries are required. The battery at the upper limit of 2·17 V per cell operates under float trickle charge conditions. This is, however, of little value in replacing capacity lost on discharge, and the battery must be removed at suitable intervals for charging.

An illustration of a battery used with a typical automatic telephone exchange is shown in Fig. 7.7.

REPEATER STATION BATTERIES

Repeater stations are connected at various points along the route of long-distance telephone systems to amplify all trunks calls. A call from a hundred or more miles away is often clearer than a local call, because a trunk call may be amplified several times *en route* from caller to receiver. Batteries installed in repeater stations may be of large capacity to supply the continuous current required by the amplifier. As in telephone exchanges the battery works in parallel with mains-operated rectifiers, so that there is continuity of supply in the event of failure of the mains.

Emergency Lighting

Storage batteries have been used for the past 30 years for emergency lighting of hospitals, public buildings, large department stores, factories, flats, banks and places of entertainment, and including more recently football grounds where floodlighting is installed.

FIG. 7.7. TELEPHONE EXCHANGE BATTERY

A typical assembly of large Planté cells of 10,000-Ah capacity

The storage battery system can claim several advantages over other lighting methods using either gas or engine-generators. It maintains a continuous supply of electricity automatically and without noise, fumes or risk of fire.

The lead-acid Planté battery, which is used extensively for this duty, is particularly suitable and can be maintained in a healthy condition by trickle charging from a simple rectifier charging unit.

PLACES OF ENTERTAINMENT

Perhaps the largest application of emergency lighting by battery is in the entertainment field, where crowds of people gather in one building.

The safety of the public in cinemas by the provision of adequate exits and an emergency lighting system is covered by Regulations made under the Cinematograph Act, 1909. These Regulations set out the various methods of illumination to be provided. They contain a strict specification relating to the size and operation of batteries, together with routine test precautions to be taken before the public are admitted to a performance. The Regulations are issued by the Home Office, but their interpretation is left to the local authority which grants a licence to the cinema. Many local authorities insist on the same standard of safety in theatres, dance halls, hotels and restaurants before a licence can be granted.

HOSPITALS AND NURSING HOMES

Emergency lighting is particularly vital in hospitals and nursing homes, not only in the operating theatres but also in the wards and corridors.

Private nursing homes in London are subject to inspection and licence by the London County Council, who wisely insist on emergency lighting in operating theatres. There is, however, no statutory requirement for emergency lighting in these places, although its importance is recognized, and many hospitals have at least a battery-operated emergency lighting system in the operating theatre.

The emergency system is arranged so as to ensure a light over the operating table in the event of mains failure, blowing of a fuse, or failure of the main lamp filament. This precaution would appear to be most essential, particularly in view of the experience of recent severe winters when during protracted cold spells the demand for electricity exceeded the generating capacity of our power stations, resulting in either widespread breakdown in supply or power cuts.

It is likely that, if funds were available, emergency lighting of operating theatres, wards and key points of all hospitals would be undertaken.

Emergency Battery and Equipment

Of the lead-acid batteries, the Planté type has been used almost exclusively for emergency lighting systems and has given a trouble-free life in excess of 20 years when operated under trickle-charge conditions.

The system consists of a battery operating in conjunction with a control cubicle housing a transformer, smoothing choke, rectifiers, switches and instruments. No special battery room is required, and when fairly small the battery can be housed along with the control equipment in the metal cubicle (Fig. 7.3).

In the larger installations the battery is carried on a wood stand adjacent to the control cubicle (Fig. 7.4).

There are two main methods of controlling the emergency-lighting battery system: floating battery scheme, and automatic switching.

FLOATING BATTERY

In the floating battery scheme, the power supply for the emergency lights normally comes from the mains through a transformer and rectifier with the battery floating across the rectifier output. In the event of a mains failure or power cut, the battery immediately takes over the load. Following an emergency discharge the battery is brought to the fully charged condition by switching to the "charge" position. Charging is terminated manually. The sequence is restored by switching the battery on float across the load. Should the emergency system be out of use for any length of time, the battery can be maintained fully charged by switching to the "trickle charge" position.

Emergency lighting battery systems with floating battery control are not widely used, as absolute continuity of supply is not essential. A momentary break of a fraction of a second can be tolerated in lighting circuits, so that a well-designed automatic change-over switch can be used.

AUTOMATIC SWITCHING

Maintained Circuit

In this method, the emergency lighting load is normally supplied from the secondary of the lighting transformer, but on mains failure an automatic change-over contactor instantly transfers the load from

the lighting transformer to the battery. Generally, all emergency
lighting change-over contactors used in premises covered by the
Cinematograph Act must comply with British Standard No. 764.
The coil of the change-over contactor (or automatic switch) is
energized from the mains, so that the battery is isolated from the

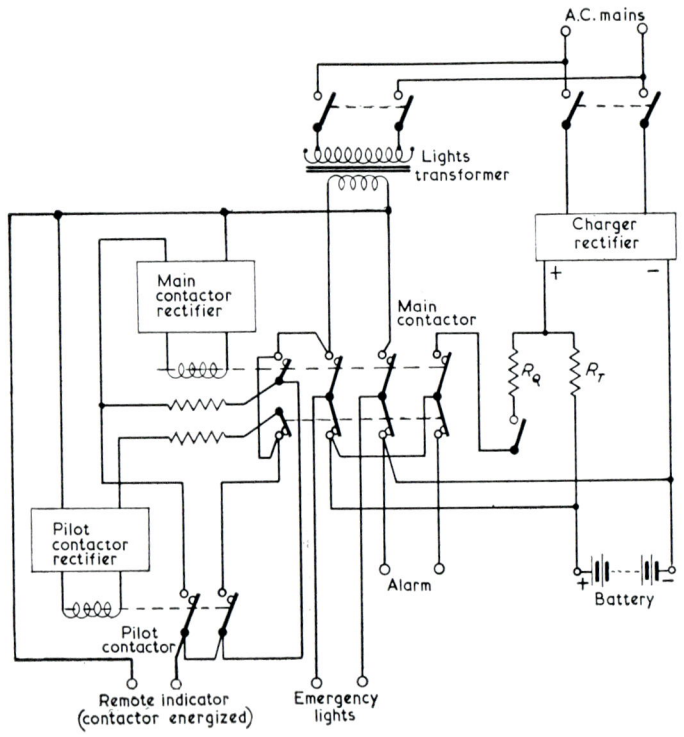

FIG. 7.8. MAINTAINED EMERGENCY LIGHTING SYSTEM

R_Q Quick-charger resistor
R_T Trickle-charge resistor

lighting circuit. When the main supply fails the coil is de-energized,
releasing the switch contacts, which close under the influence of
gravity or springs, and immediately connect the battery to the load.

Following any emergency discharge, the battery is fully charged by
setting the charge switch to the "quick charge" position. At other
times the battery is maintained healthy and fully charged by a
continuous trickle charge.

This maintained circuit is suitable for safety lighting under the

Cinematograph Regulations effective from January, 1956, and where
it is a statutory requirement that the safety lights must be on whilst
the public are in the building.

The battery and emergency lighting voltage is not necessarily equal
to the mains voltage, as with a.c. mains it is possible to use a step-
down lighting transformer to feed low-voltage lamps.

Fig. 7.8 shows a typical arrangement for large or small lighting

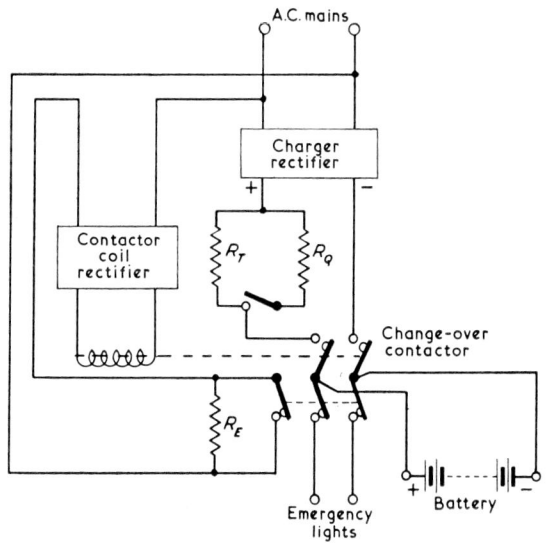

FIG. 7.9. NON-MAINTAINED EMERGENCY LIGHTING SYSTEM

R_Q Trickle-charge resistor
R_T Quick-charge resistor
R_E Economy resistor

loads. It will be noticed that there is a separate emergency-lights
switch to enable the safety lights to be switched off when the cinema
is closed.

Non-maintained Circuit

With this method the emergency lights are normally off. During a
mains supply interruption, the emergency lamps are immediately
switched across the battery by the automatic change-over contactor.
This system is particularly suitable for hospitals, department stores,
offices, factories, and places which are in use 24 hr a day. Fig. 7.9 is
a typical circuit diagram.

Battery and Equipment Maintenance

FLOATING-BATTERY INSTALLATIONS

1. Readings of total battery voltage and the specific gravities of a few cells selected as "pilots" should be recorded daily, so as to check for possible abnormalities in battery condition.

2. Freshening charges should be given weekly until the battery voltage and the specific gravities of the "pilot" cells show no further rise during a period of one hour.

3. Every month, prior to the commencement of the fourth freshening charge, the level of the electrolyte should be checked and pure water added where necessary. At the completion of the charge, the specific gravity of each cell should be recorded.

4. Any cell showing a reading noticeably lower than the other cells should be examined for internal short-circuits such as may be caused by small pieces of scale bridging across the plates.

5. Following any emergency or accidental discharge, and after any test discharge, the battery should be recharged. This charge should be continued until readings of battery voltage and specific gravity show no further rise during a period of one hour.

6. The floating voltage should be checked. If set too high it will cause appreciable gassing of the plates, and topping-up may be required frequently. If set too low the plates will have an unhealthy appearance, and the specific gravity will be below the fully charged value.

7. The battery and surroundings should be maintained dry and clean and all connexions tight.

BATTERIES MAINTAINED BY TRICKLE CHARGE

Provided that the battery is fully charged before being placed on trickle charge, and if no discharge occurs, a continuous trickle charge will keep the battery fully charged and healthy.

If the cells gas noticeably and voltages exceed 2·35 V per cell, the trickle charge should be reduced slightly. Any change made in the trickle-charge rate should be fairly small and the effect observed over a period of several months. After any discharge other than a momentary test, the battery must be recharged at the "quick charge" setting. On completion, the quick charge is discontinued and the battery put on trickle charge. Trickle charging will not recharge a discharged battery.

READINGS

Readings should be taken at weekly and monthly intervals as described for floating battery installations.

Whether operated by the float or trickle charge method, batteries do *not* require cycles of discharge and charge to maintain them in a healthy condition.

TESTING THE CHANGE-OVER CONTACTOR

With the battery operating normally on trickle charge, the "mains" switch is turned off. The contactor should then automatically transfer the emergency lights to the battery.

Inspection

The person responsible for the electrical plant can usually carry out routine maintenance of battery and equipment.

It is advisable to have a detailed inspection of the plant by some qualified person from outside, at about 6-monthly intervals. This ensures that the battery and gear are being maintained correctly. This service inspection can best be carried out by a representative of the battery manufacturer who supplied the complete equipment.

CHAPTER 8

ALKALINE BATTERIES

IN the early days of the development of storage batteries, scientists were experimenting with electrodes and electrolytes other than lead and sulphuric acid, in their search for the ideal reversible cell. Most of these experiments failed to produce a cell having the commercial value of the lead-acid couple, with the exception of the nickel-alkaline cells invented by Jungner, of Sweden, in 1899, and Edison, of America, in 1900.

The modern nickel-cadmium and nickel-iron cells with caustic (alkaline) electrolyte are developments of the types patented by these two scientists. The present nickel-cadmium battery has been developed and designed to be suitable for all battery applications, portable and stationary. The nickel-iron battery is particularly suitable for heavy traction work.

Construction

NICKEL-CADMIUM CELLS

Both positive and negative plates are built up of pockets of finely perforated ribbons of steel which totally enclose the active materials. The active material for the positive plate is nickel hydroxide, and for the negative plate a mixture of cadmium and iron.

The active materials are formed into briquettes under high pressure and fed between perforated steel ribbon, which is double folded at the edges to produce a continuous strip filled with active material.

The double fold along the edge of the strip provides a recessed lip, and the long strips are placed side by side so that the lip of one strip fits into the recess of the adjacent strip.

As many strips as may be necessary to produce the desired length of plate are interlinked and pressed into a uniform flat plaque by passing between rollers.

The plaque is slid into a plate frame of two channelled steel side frames welded to a top bar. The complete plate (Fig. 8.1) is subjected to high pressure, which corrugates the surface and indents the vertical grooves which locate the ebonite rod separators in position.

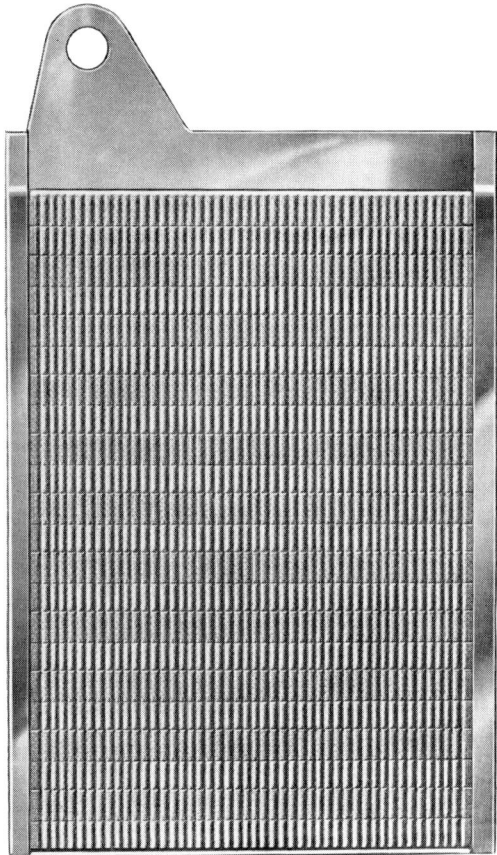

FIG. 8.1. NICKEL-CADMIUM PLATE
(Nife Batteries)

NICKEL-IRON CELLS

The negative plate is generally similar in construction to those
described above, but the active material encased in the steel pockets
is iron oxide.

The positive plates consist of nickel-plated perforated steel tubes
reinforced by seamless steel rings. The tubes are packed tightly with
the active material of nickel hydrate and layers of pure nickel flake to
improve conductivity. The positive plate (Fig. 8.2) is completed by
clamping a number of tubes in a strong steel frame.

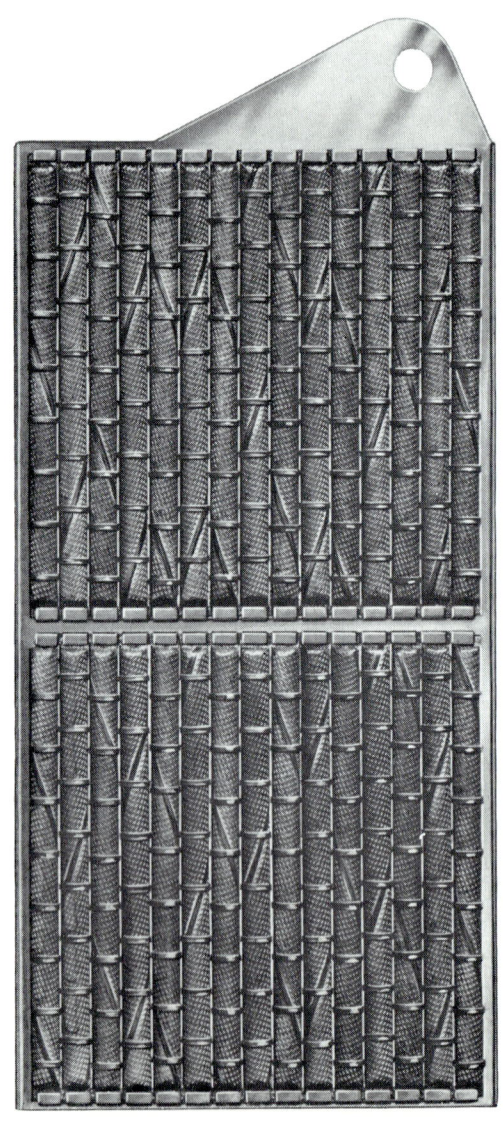

FIG. 8.2. TUBULAR POSITIVE PLATE OF NICKEL-IRON CELL
(*Britannia Batteries Ltd.*)

FIG. 8.3. NICKEL-CADMIUM ELEMENT ASSEMBLY
(*Nife Batteries*)

PLATE GROUPING

Plates of similar polarity are threaded on to a steel assembly bolt, with steel spacing washers between each pair of plates. In the centre of each assembly bolt is fitted a steel terminal pole.

Positive and negative plates are interleaved and insulated from each other by thin ebonite rod separators (Fig. 8.3).

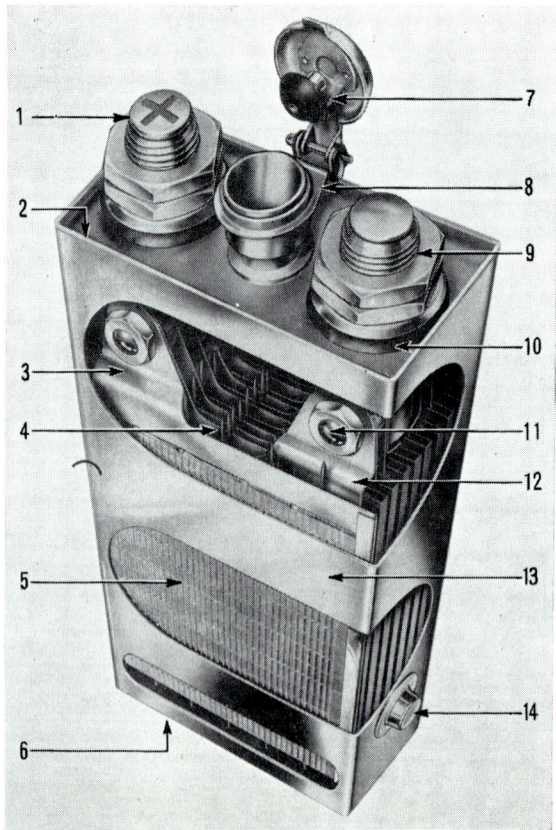

FIG. 8.4. NICKEL-CADMIUM CELL

(Britannia Batteries Ltd.)

1. Positive terminal
2. Cell cover (welded joints)
3. Positive plate frame
4. Insulating rods
5. Positive plate
6. Steel bottom (welded joints)
7. Vent cap
8. Filler cap
9. Negative terminal
10. Liquid-tight gland
11. Steel assembly bolt
12. Negative plate frame
13. Welded steel container (protective finish)
14. Steel suspension boss

CONTAINERS AND CELLS ASSEMBLY

The cell containers, of welded construction, are made from steel sheets, and are cadmium and nickel plated to prevent corrosion of the outer surface. Steel suspension bosses, for assembly in wood crates, are welded on to the sides of the container. The container lid carries a combined vent and filling orifice; on large cells this is

FIG. 8.5. FIVE-CELL NICKEL-CADMIUM BATTERY IN WOOD TRAY
(Britannia Batteries Ltd.)

welded to the lid and contains a non-return poppet valve—on small cells a screw-in plastic vent plug is used. The cell terminal posts pass through the lid via leak-proof stuffing boxes sealed with rubber glands. The terminal posts are threaded to take nuts which securely hold the intercell connectors in contact with the post.

A complete cell assembly with cut-away section is shown in Fig. 8.4.

Because the steel containers are electrically "alive," being in contact with the conducting electrolyte (potassium hydroxide solution), each cell must be insulated from its adjacent cell and from earth. This is achieved, as shown in Fig. 8.5, by suspending each cell by the bosses welded on the container sides. The boss fits into a corresponding recess in the framework of the wood tray.

Chemical Reactions

The chemical reactions of the alkaline battery are rather complex, but expressed simply, consist of the transfer of oxygen from one plate to the other. The electrolyte, consisting mainly of potassium and lithium hydrates diluted with distilled water to about 1·210 sp. gr., conducts the current by ionization but does not react with either plate. The specific gravity, therefore, does not change materially on either discharge or charge.

During discharge, oxygen is transferred from the positive to the negative plate, and during charge is retransferred to the positive. Beyond this there is no general agreement about the reactions in this cell, but for practical purposes they are sometimes written

$$\overset{Charged}{\underset{\substack{\text{Nickel hydrate} \quad \text{Cadmium} \\ \text{Positive} \qquad \text{Negative}}}{2Ni(OH)_3 + Cd}} \rightleftarrows \overset{Discharged}{2Ni(OH)_2 + Cd(OH)_2}$$

Nickel-Cadmium Cells (Normal Resistance)

A range of nickel-cadmium cells of "normal resistance" has been designed for a wide variety of applications, including stationary and standby duties, emergency lighting, electric traction, and many portable duties.

Where the full capacity of the battery is regularly required as in traction and similar applications, a special double-pocket positive plate is used. This double set of pockets in the single frame enables the maximum quantity of active material to be used. The negative plate is of a special single-pocket construction to match the capacity of the positive plates.

Extra-low-resistance Cells

The construction of the extra-low-resistance cell is basically the same as that of the normal-resistance type, but a larger number of thin plates, assembled closer together, is used.

An extra-low-resistance cell of 100 Ah would have 23 plates compared to 11 plates for the 100-Ah normal-resistance cell. The increase in plate area and reduced internal resistance make this cell particularly suitable for applications involving very heavy current, short-duration discharges.

Characteristics

Nickel-alkaline batteries have certain characteristics which make them particularly suitable for a wide range of operating conditions. These include—

(*a*) Great mechanical strength against vibration, shock and abuse.

(*b*) Ability to stand idle at normal temperature for long periods in various states of charge, without deterioration and with negligible loss of charge.

(*c*) Overcharge at normal rates of charge, and occasional boost charges at high rates, do not adversely affect the battery life. (This assumes that the average operating temperature is not excessive and that the occasional maximum temperature does not exceed 46° C (115° F).)

(*d*) For traction work, regular recharging in 7 hr is possible and advisable in the interests of maintaining the charging current above a recommended useful minimum.

(*e*) Long life (8 years or more) when maintained correctly.

Discharge and Charge Voltages per Cell

	Nickel-cadmium	Nickel-iron
Discharge	V	V
Normal rates (2–10 hr)	1·2	1·2 (approx.)
Engine-starting rates	0·8 (approx.)	Not used
Charge		
Normal rates (7–10 hr): start . .	1·4	1·5
finish . .	1·7	1·84

Because of the lower discharge voltage, more cells must be used for a given system voltage when using alkaline cells than when using lead-acid cells. For example,

	Nickel-alkaline	Lead-acid
Engine-starting 24-V system . . .	18–20 cells	12 cells
Engine-starting 110-V system . . .	72 cells	48 cells
Switch-closing 110-V system . . .	92 cells	55 cells

The physical and electrical characteristics of the alkaline battery enable it to operate successfully under the conditions of abuse that may be present in many portable applications. A complete range of nickel-cadmium batteries is available for handlamps, portable lighting sets, ships' emergency lanterns, laboratory work and military purposes, in addition to the larger applications given in greater detail below.

Traction

Traction batteries are operated on regular cycles of discharge and charge, and both nickel-iron and nickel-cadmium types are used.

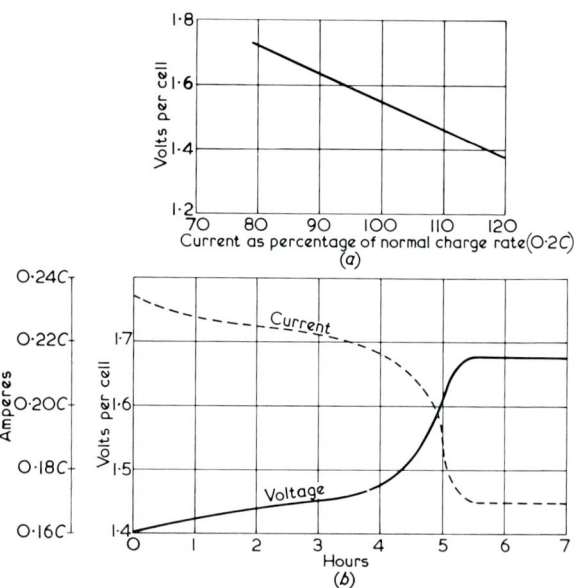

FIG. 8.6. TYPICAL RECHARGE CHARACTERISTICS OF A NICKEL-CADMIUM BATTERY FROM A TAPER CHARGER

(*a*) Rectifier voltage/current characteristic
(*b*) Current/time and voltage/time characteristics
C = Capacity at 5-hr rate

As the batteries are usually substantially discharged on each working cycle, it is essential to give a generous recharge to maintain all cells in a healthy condition. For every 100 Ah taken out on discharge, 140 Ah should be put back on recharge, and the rate of charge should not be less than two-thirds of normal. The normal rate in amperes is one-fifth of the nominal ampere-hour capacity.

For example, following the discharge of a 100-Ah battery, the recharge should be 20 A for 7 hr, or 14 A for 10 hr.

In traction service, single-step taper chargers are used extensively, but the rectifier characteristic differs appreciably from that used with lead-acid batteries (Fig. 8.6 (*a*)).

Fig. 8.6 (*b*) shows the voltage and current characteristics of a nickel-cadmium battery when recharged from a typical taper charger, Fig. 8.7 shows a set of charge characteristics for a nickel-iron battery.

Automatic termination of charge by means of a voltage-time relay is used in charging nickel-cadmium batteries, the voltage-sensitive

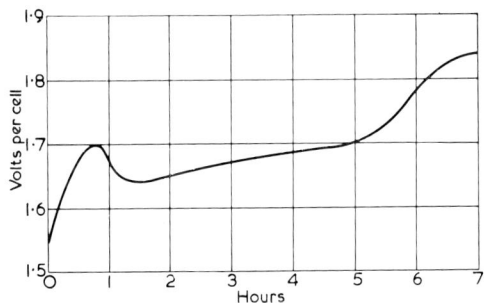

FIG. 8.7. TYPICAL VOLTAGE/TIME RECHARGE CHARACTERISTIC OF NICKEL-IRON BATTERY

Charging current = 0·2 × Capacity at 5-hr rate

relay being set to operate at a voltage equivalent to 1·6 V per cell, and the timing motor set to run for approximately 2 hr. It will be seen from these curves that the voltage on charge of a nickel-iron battery differs from that of a nickel-cadmium battery. An increase in the nickel-iron battery voltage, up to a value of almost that attained at the end of charge, occurs early on in the charge. This initial voltage swing, early in the charge, means that the voltage-sensitive relay control method is unsatisfactory, and a process timer is used instead.

Engine Starting (Portable Applications)

Engine-starting batteries are mainly used on railways. Because the batteries have to supply very heavy discharge currents for starting Diesel engines on Diesel-electric main-line locomotives, Diesel railcars and Diesel shunters, the extra-low-resistance type is always used. Typical characteristic curves are shown in Fig. 8.8.

This type is commonly used also for emergency lighting on electrified railstock involving high rates of discharge during power failures.

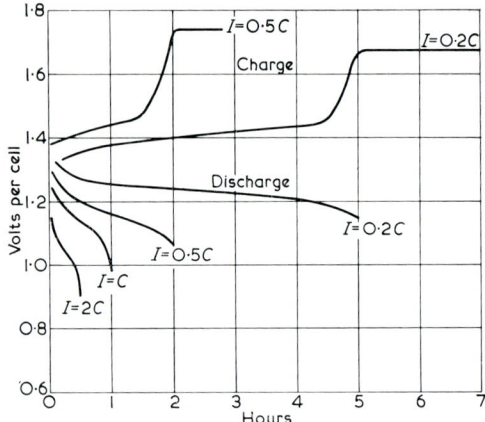

FIG. 8.8. TYPICAL VOLTAGE/TIME DISCHARGE AND CHARGE CHARACTERISTICS
OF NICKEL-CADMIUM EXTRA-LOW-RESISTANCE BATTERIES

I = Current
C = Capacity at 2-hr rate
Temperature, 20° C (68° F)

Train Lighting

Nickel-cadmium batteries of the normal-resistance type are used for general lighting duties on steam or Diesel-hauled trains where the duty is of a discharge/charge nature. These cells have the same

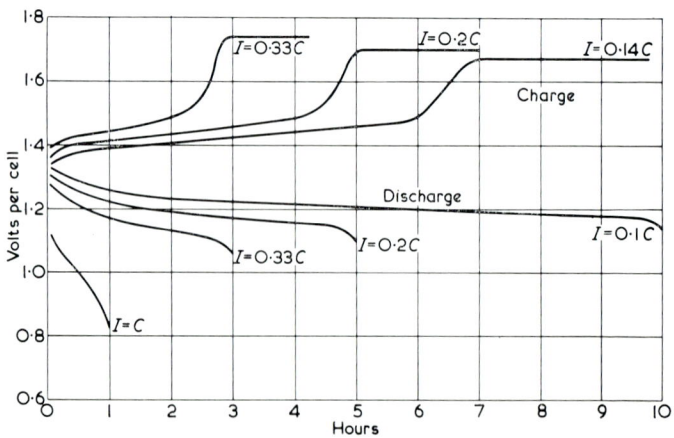

FIG. 8.9. TYPICAL VOLTAGE/TIME DISCHARGE AND CHARGE CHARAC-
TERISTICS OF NICKEL-CADMIUM NORMAL-RESISTANCE BATTERIES

I = Current
C = Capacity at 10-hr rate
Temperature, 20° C (68° F)

electrical characteristics as standard-type cells, but there is provision for a much larger volume of electrolyte above and around the plates to lengthen the interval after each topping up operation. In addition to a longer life expectancy, these batteries are of considerably less weight and volume compared to a lead-acid battery of similar capacity. Typical characteristic curves are shown in Fig. 8.9.

Railway Signalling

The nickel-cadmium battery has been used extensively on automatic train control systems, and has been adopted for use on the British Railways automatic warning system. A battery mounted on the footplate of the locomotive provides an audible signal to the driver when passing signals as to whether they are on or off, and a relay operated by the batteries applies the brakes automatically when necessary.

Stationary Applications

The nickel-cadmium battery is used for almost all the stationary applications described in detail in Chapter 7. These include switch-gear operation (closing and tripping), and standby and emergency duties.

For switch closing, the extra-low-resistance battery is used and the number of cells is determined by the system voltage. For this duty the nominal cell voltage of approximately 1·2 V would demand 92 cells for a 110-V system, or 200 cells for a 240-V system.

The capacity of the battery is determined by the load and the minimum desirable battery voltage equivalent to 1 V per cell. This ensures that the voltage at the closing solenoid is not less than 80 per cent of nominal value.

Tripping batteries are of the normal-resistance type, as the load current in small substations is relatively light and intermittent. In many cases the battery is small enough to be housed with the charger in a steel cubicle. A typical self-contained tripping unit is shown in Fig. 8.10.

Power station batteries vary considerably in capacity and voltage, and are usually very much larger than main substation batteries. Where the load is essentially emergency lighting and oil-pump motors, with switch closing as a secondary load, the battery is of the normal-resistance type.

For standby and emergency lighting duties in most places of entertainment, large stores, hospitals, etc., the normal-resistance type of battery is used. Depending on the loading and the battery size required, the installation may either be a self-contained unit, where

FIG. 8.10. SWITCH-TRIPPING UNIT USING NICKEL-CADMIUM NORMAL-
RESISTANCE BATTERIES

(*Nife Batteries*)

the battery and charger are housed together in a steel cubicle, or a major unit control cubicle with the battery mounted separately on a single-tier stand.

Other important applications are marine, and for engine starting on standby sets.

The alkaline battery withstands extremely well the arduous operating conditions of marine service. Where vessels, yachts or power boats are laid up for long periods, the alkaline battery can be left without any special attention. It is, however, preferable to leave it in a charged condition and to give an extended charge before it is put back into service.

When it is used solely for engine starting on standby emergency power plant, the absence of self-discharge enables the alkaline battery to function after long stand periods. Charging is only required after a stand period of 6 to 12 months, and following a number of engine starts. Where the battery is required intermittently for regular duties, it is necessary to give more frequent charges either by the engine-driven generator or by a static charger if a mains supply is available.

Operating and Maintenance Instructions

1. Avoid over-discharging, and ensure that charging is adequate for the particular duty.

2. Maintain the electrolyte at the correct level by topping up with pure distilled water only. Excessive consumption of water indicates overcharging or battery operation at too high a temperature. Negligible consumption of water might indicate that the battery is being undercharged.

3. Keep the cells and trays clean and dry.

4. On no account use utensils which have contained acid. *Sulphuric acid will destroy any alkaline battery.*

5. Terminals, connectors and tops of cell containers should be lightly smeared with petroleum jelly.

6. Renew the electrolyte when the specific gravity falls to 1·160.

Safety Precautions

Do not allow metal objects to rest across or fall between the cell containers, as they are "alive."

Keep naked flames away from the battery, as with all storage batteries the gases evolved during charging are explosive. Avoid contact with the electrolyte either by the skin or clothing.

CHAPTER 9

MISCELLANEOUS TYPES

THIS chapter describes various batteries designed to meet rather special applications not covered in previous chapters.

Aircraft Batteries

Batteries are an essential part of the electrical equipment of all aircraft, as they maintain vital loads when the main generator is stopped. For example, in an emergency a battery would provide

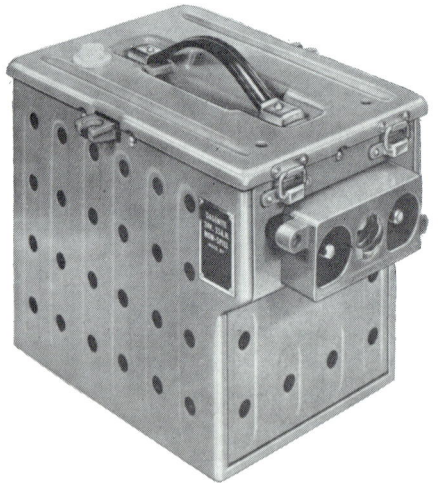

FIG. 9.1. 24-V 25-AH AIRCRAFT BATTERY

electrical power for the operation of the v.h.f. radio equipment required to make a safe descent.

Aircraft batteries are usually made as 6-cell (12 V) or 12-cell (24-V) units; a typical 24-V 25 Ah battery is shown in Fig. 9.1. This type of battery is suitable for use on the Viscount, Vanguard, Comet IV and many other aircraft, and like most aircraft batteries is designed to have a high capacity/weight ratio particularly at high discharge rates for engine starting.

The construction consists of cells in high-impact polystyrene containers housed in a pressed aluminium box with an aluminium top cover seating on the box to form a ventilation chamber. The top cover is secured by toggle catches, and has holes for the admission of air and a polystyrene nozzle for connexion to a pipe. The pipe enables gas and vapours given off during charge to be removed to a safe place, such as outside the aircraft. This arrangement gives positive ventilation and avoids the provision of a specially ventilated battery bay in the aircraft.

MAINTENANCE

Most aircraft batteries are designed to be fully or semi-unspillable, which property depends on the level of the electrolyte being at the correct height above the separators. Acid levels should be checked in every cell at least once a fortnight on the aircraft and also at every bench charge. Low levels should be corrected by adding the right volume of pure water.

The electrolyte specific gravity of all cells should be checked at fortnightly intervals when the battery is in service, and if below 1·230 (corrected to 15° C), the battery should be given a bench charge.

BENCH CHARGES

Batteries should be charged at the bench charge rate recommended by the battery maker until voltages and specific gravities cease to rise for 1 hr. If necessary, top up with pure water before charging.

CAPACITY TEST

If low capacity is suspected, and in any case once in every three months, a capacity test should be taken. The battery should first be fully charged and then discharged by connecting to a suitable resistance with an ammeter in circuit. A constant discharge current at the 10-hr rate should be maintained until the battery voltage has fallen to the equivalent of 1·8 V per cell (10·8 V and 21·6 V for 12-V and 24-V battery, respectively). If the discharge period is 8 hr or more, the battery should be recharged and returned to service. If it is less than 8 hr the battery should be given an extended charge and the capacity test repeated.

If the battery again fails to give 8 hr it should be considered unserviceable.

INSULATION TEST

An insulation test is usually carried out by connecting a 250-V Megger between one terminal of the battery and a metal plate on which the battery is standing. A figure of 1 MΩ should be obtained with a perfectly clean and dry battery. If a figure higher than 10,000 Ω cannot be obtained, it is possible that a cell is leaking or the battery is not perfectly dry. To check this point the battery should be washed down with water and thoroughly dried in a current of warm air, and the insulation test repeated.

GENERAL

When the battery is out of service it should be maintained fully charged by giving a freshening charge every 4 to 6 weeks.

Holes in vent plugs should be checked for free passage by blowing through the plug. Vent plugs should be well screwed home at all times.

The battery should be kept clean and dry, and the terminals coated with a thin layer of petroleum jelly.

Cap-lamp Batteries

Battery-operated cap lamps have have been used for more than 40 years in coal mines. The equipment consists of a lead-acid battery fastened to the miner's belt and a lamp head-piece, containing the bulbs, fixed in the helmet. A two-core flexible cable connects the lamp head-piece to the battery.

A 2-cell (4-V) lead-acid cap-lamp battery is shown in Fig. 9.2. This is capable of supplying current to light a lamp bulb for 9 hr.

Each cell has one positive plate and two negative plates. The positive plate is of Gauntlet construction using a sheath of woven Terylene. The pasted-type negative plates are enclosed in a sleeve of microporous p.v.c. to eliminate the possibility of internal short-circuits. The battery is completely leak-proof and virtually unspillable in any position, owing to a special venting and sealing system and a restricted amount of free acid. Most of the acid is soaked up by the plates and a composite form of separator. This consists of a glass-felt mat in combination with a highly absorbent separator made from treated wood fibres, together with a Porvic envelope enclosing the negative plate.

The battery container is a two-compartment moulding of special-grade hard rubber designed to allow for abrasive wear in service.

SELF-SERVICE OPERATION

For some time it has been the practice for the miner to collect his charged battery and lamp direct from the lamp room at the start of his shift, and to put the battery on charge on his return.

The charging equipment consists of large rectifier chargers supplying 5 V d.c. to rows of charging racks. There is therefore no danger of electric shock, and the battery is protected against overcharging. Charging is by the modified constant-voltage method in which the

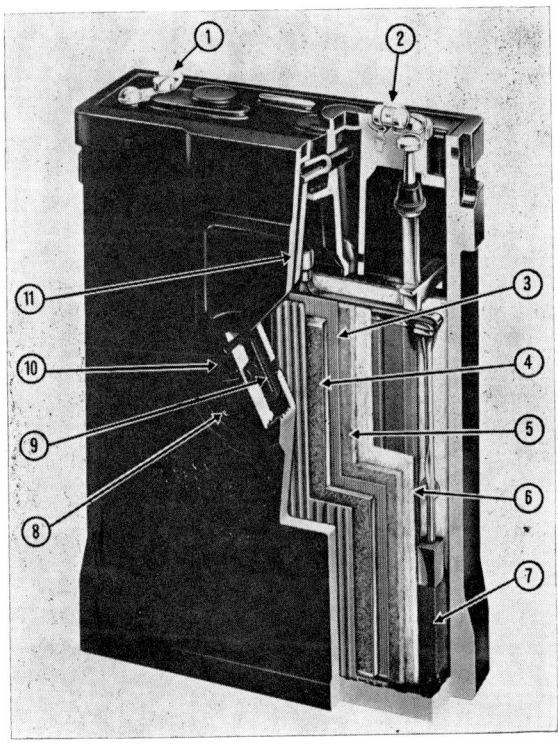

FIG. 9.2. MINER'S CAP-LAMP BATTERY

1. Polarized cable-ends
2. Replaceable fuse unit
3. Porvic envelope
4. Pasted negative plate
5. Absorbent Lignex separator
6. Absorbent glass-wool separator
7. "Gauntlet" positive plate
8. Keyhole
9. Rubber disc with captive anti-friction washer
10. Screwed vent plug
11. Vertical gas passage

current falls to a safe value as the battery voltage rises on charge. The current at the commencement of charge with a fully discharged battery is limited by including a ballast resistance of about $0.3\ \Omega$ in the charging leads to each battery.

MAINTENANCE

Batteries and lamps are maintained by the lamp-room staff. Batteries are regularly topped up, washed and cleaned. Lamps are examined for damage, dirty reflectors and contacts, and bulbs are renewed at regular intervals to avoid lamp failures down the mine.

Diesel-Engine Starting Batteries

The starting of heavy Diesel engines of the type used in Diesel-electric standby sets, multiple train units, or long-distance Diesel-electric locomotives requires a battery capable of giving very high

FIG. 9.3. FOUR-CELL DIESEL-ENGINE STARTING BATTERY

discharge rates. In fact this type has a greater starting performance in relation to its nominal capacity than any other battery of the same capacity.

The plates are generally similar in design to those used for the heavy-duty automotive batteries, except that they are thicker and more robust to withstand the vibration and physical shocks of railway service. Dual separators of microporous plastic sheets and glass-wool mats fitted between the plates provide a compact assembly and an effective retainer of the active material of the positive plates.

Batteries similar to the unit shown in Fig. 9.3 are made of separate cells assembled in hardwood trays or monobloc hard-rubber containers.

The largest batteries used for starting the engines of main-line Diesel-electric locomotives usually consist of 48 cells working across a nominal 110-V supply. The batteries must be capable of supplying power to motor the main generator on starting, at a temperature as low as 0° C and in a half-discharged condition.

The batteries normally work "on balance" in conjunction with a constant-voltage generator driven by the engine. The generator is voltage regulated to provide a voltage equivalent to 2·2–2·3 V per cell, depending on the particular operating conditions. Under most operating conditions the batteries are maintained substantially fully charged.

Specific gravity readings should be taken at about monthly intervals, and if below about 1·200 the battery should be given a bench charge until voltages and specific gravities are constant for 3 one-hourly readings. This supplementary charging of a discharged battery is necessary as the system voltage is too low to restore cells to the healthy, charged condition.

Low-Loss Batteries

Low-loss batteries use plates specially designed to reduce internal losses to about one-seventh of those of the ordinary pasted-type plate. The plate grids are made from a low-antimony or pure-lead alloy to prevent antimony transfer to the negative plate during service. When deposited on the negative plate, antimony produces local action and self-discharge of the plates. This is not important for batteries which are regularly charged, but where a charge has to be retained for periods up to 12 months it is necessary to use the low-loss type of battery. These are used for light-current work, usually of an intermittent nature and frequently with long periods on open-circuit between discharges.

Batteries in the low-loss class, shown in Fig. 9.4, range from the single 2-V unit with thick plates spaced well apart in a translucent polystyrene or glass container to the compact 6-V unit of three cells in a monobloc polystyrene container. They are used for a wide variety of duties including bells, clocks, fire alarms, electric fences, and laboratory work.

Fig. 9.4. Low-Loss Range of Cells and Batteries

When used with mains systems, as in some clock and fire-alarm installations, the batteries can be maintained indefinitely by a trickle charge from a mains-operated rectifier charger. In many applications, however, there is no provision for recharging, and the battery may be left connected to the equipment until it fails to operate. The battery then becomes severely over-discharged, and repeated working under such conditions will shorten its life. In the interests of long battery life and reliability it is advisable to recharge at least every 6 months, or whenever the specific gravity falls to 1·180, whichever occurs first.

Marine Batteries

Large passenger liners are to some extent floating power stations as so much of their equipment requires electrical power, not only for navigation and radio communication but also for the general comfort of the passengers.

As in land power stations, failure of electrical power must be catered for by providing a standby supply. In most cases this supply is provided by storage batteries.

The International Conference on Safety of Life at Sea has recommended that all passenger ships and large cargo vessels should have an emergency lighting and power system supplied by a battery together with an engine-generator set which can take over the emergency load from the battery after half an hour.

Large emergency battery equipments provide power for boat stations, watertight doors and emergency steering, as well as lighting for alleyways, stairways and navigation. A battery of mains voltage can be divided into two equal sections for trickle charging and quick charging from d.c. mains. When used with a.c. mains the battery is charged from a suitable rectifier charger.

Smaller batteries of 25 or 50 V can be installed in the engine room and charged from the ship's main supply. These batteries supply emergency lights, which are automatically switched on when the main supply fails.

Batteries used for marine work may be of either the lead-acid or the nickel-alkaline type.

The lead-acid marine battery is of the armoured-plate type of construction, generally similar to that used for traction batteries as described in Chapter 5. The plates may be either

    (*a*) Tubular plates of Exide Ironclad Gauntlet construction.
    (*b*) Heavy flat plates.

Fig. 9.5. Part of a Flat-Plate Marine-type Battery

Individual cells are assembled together in wood trays fitted with lifting irons and side insulators. Part of a battery is illustrated in Fig. 9.5.

Marine Radio Batteries

One important item of ship's equipment which requires an electrical supply independent of ship's mains is the radio transmitter and receiver.

At one time standard car batteries were used for this duty, but these have been superseded by a battery using a more robust design incorporating double separation of glass-wool mats and microporous p.v.c. separators.

For many purposes it is customary to use two 24-V batteries of 128 Ah capacity at the 10-hr rate, so that one battery is operating the set whilst the other is on charge.

OPERATING INSTRUCTIONS

The following recommendations should be followed to obtain maximum battery life.

1. Discharging

Do not regularly discharge the battery beyond the 60 per cent discharged condition, corresponding to about 1·180 sp. gr. reading. The reserve of capacity should not be used except in an emergency.

2. Recharging

Recharge the discharged battery at the recommended normal rate (5–7 A per 100 Ah), until the cells commence to gas and the specific gravity is about 10 points (0·010) below the fully charged value. This will avoid overcharging.

3. Equalizing Charge

After every seven cycles of discharge and charge the battery should be given an extended charge. This is done by continuing an ordinary charge for a further 2 hr. At the end of this period, specific gravity readings should be recorded at hourly intervals until they are constant for 3 hr.

4. Topping-up

Distilled or pure water should be used. (If supplies fail it is better to use drinking water than to allow the acid level to fall too low.)

Batteries should be topped up to the correct level at least once a month.

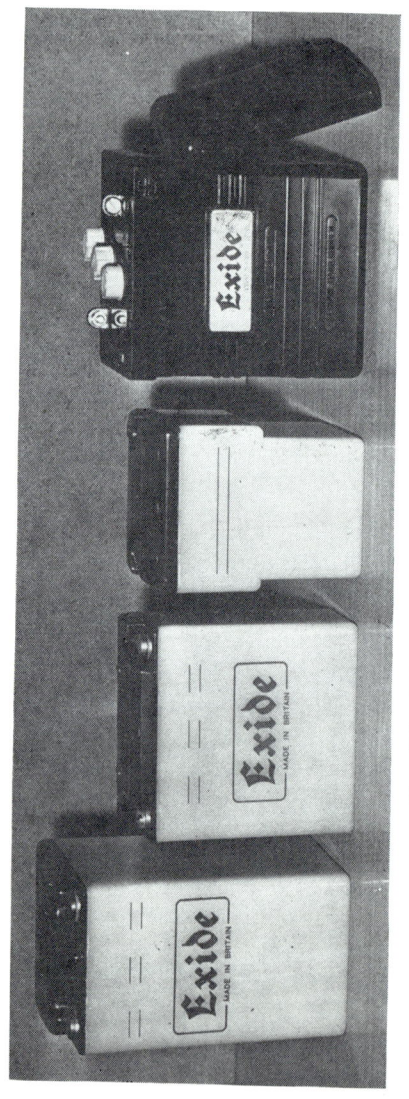

Fig. 9.6. Scooter and Motor Cycle Batteries

5. *Idle Batteries*

It is unnecessary to give a battery artificial cycles when not in use. It is advisable to charge an idle battery at least once every 3 months, or once a month if it is more than a year old, to replace standing losses.

6. *Maintenance and Safety Precautions*

Tops of batteries should be kept clean and dry, and terminals coated with petroleum jelly.

The usual safety precautions against explosions should be observed, that is, keep a flame or spark away from the battery particularly during and after a charge.

Motor-Cycle and Scooter Batteries

There is a wide range of small, light-weight batteries available for the various types and sizes of motor cycle, scooter and three-wheel machine.

The latest batteries, some of which are shown in Fig. 9.6, use polystyrene containers, with plates designed to give the high starting performance required for machines fitted with electric starting.

The batteries are made as 6-V units, with a capacity ranging from 5 Ah to 25 Ah. The largest sizes are used for three-wheelers and radio-equipped motor cycles. For a machine with electric starting, a 12-V battery consisting of two 6-V units in series is used.

Portable Emergency Lighting Batteries

The equipment shown in Fig. 9.7 is designed for use where permanent emergency lighting schemes (*see* Chapter 7) are impracticable. It is particularly suitable for installations requiring one or two emergency lighting points only. It simply plugs in to any power point, and no special wiring is required. The battery and charger are housed in separate compartments of the steel container.

The 12-V battery is designed to have low-loss characteristics and to be suitable for trickle charging, and is capable of supplying a single 36-W lamp for $3\frac{1}{2}$ hr, or two lamps (not exceeding 60 W total) for $1\frac{1}{2}$ hr.

The lamp, which may be connected to the unit either by plug and flexible lead or by a swivel mounting on top of the case, is automatically switched on when the a.c. supply fails.

Maintenance is negligible as the battery is kept charged from the mains charger, and topping-up is necessary only about three times per year.

Submarine Batteries

Lead-acid batteries have been used extensively in submarines since the first British submarine was launched in 1901. The batteries are used mainly to drive the electric motors for propulsion when the submarine is submerged. They also provide power for a consider-

FIG. 9.7. PORTABLE BATTERY LIGHTING SET

able amount of electrical equipment when the main Diesel-generators are not running.

The submarine cell is, in a way, a much enlarged version of the traction cell but using a large number of plates in order to give the maximum possible output per unit weight and volume. A typical cell, Fig. 9·8, weighs about 500 kg and has a capacity of approximately 7,000 Ah at the 5-hr rate.

The container and cover are of glass fibre impregnated with resin, and are specially designed to withstand severe shocks resulting from depth charges. The container is lined with a rubber bag cemented

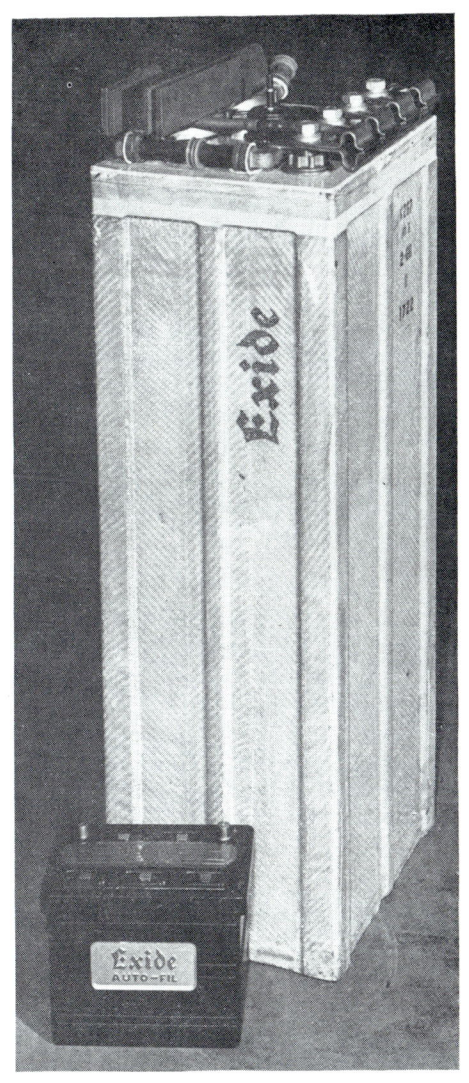

FIG. 9.8. 2-V 7,000-Ah Submarine Cell Compared for Size with
12-V 43-Ah Car Battery

Connectors of submarine cell are water cooled

to the inner walls, to prevent leakage of electrolyte from the cell in the event of damage to the container.

A submarine battery may consist of up to 448 cells, with 4 sections each of 112 cells. Sections may be connected in series or parallel as required.

Train Lighting and Air-conditioning Batteries

The armoured type of positive plate is used almost exclusively for train lighting and air-conditioning cells.

It may be either the tubular design of plate using slitted ebonite tubes or gauntlet sheaths of woven Terylene, or the heavy flat plate used with glass-wool armouring.

These designs are very similar to those used for traction cells described in Chapter 5. The train cell box differs, however, in having a much thicker wall, reinforced on the sides by vertical ebonite strips. This enables the cells to be butted up against each other without packings, and at the same time leaves a space for ventilation and cooling. The cell box, being self-supporting, does not need to be fitted into outer wood trays.

Frequency of topping-up is reduced by having ample head-room for acid above the plates, and the provision of an acid-level indicating float enables topping-up to be carried out accurately without any necessity to look into the cell.

Large batteries consisting of 56 cells for a nominal 110-V system would be charged from the axle-driven generator by the current-voltage control system. By this method the battery is charged at constant current equal to the maximum output of the generator until the battery voltage reaches the equivalent of about 2·42 V per cell, at which point the battery would be at least 80 per cent charged. Thereafter the voltage is kept constant so that the current falls to a safe value as the battery becomes charged.

Batteries for Cordless Appliances

All equipment worked by batteries carried either in the equipment or housed outside with an interconnecting cable can be classed as "cordless appliances," that is, it functions independent of a mains supply.

So far we have considered equipment ranging from the giant submarines whose batteries weigh 200,000 kg (200 tons) down to small cap lamps with a battery of a kilogramme (2 lb) or so. In recent years an increasing number of consumer appliances have come to be known as "cordless appliances." They include hearing aids, transistor radios, portable television sets, tape recorders, cameras, toys, electric shavers, hedge trimmers and lawn mowers. Others such as electric tooth brushes, carving knives and fish-bite line indicators are more common in the United States than here in the United Kingdom.

Torches, small handlamps and lanterns and some of the above appliances require relatively little power from the battery and the small current drain can be supplied quite adequately by the well-known dry battery (carbon-zinc, Leclanché type). In fact for low power appliances this battery is the best at the price.

The history of the dry carbon-zinc battery goes back as far as that of the lead-acid battery, at which time it was used mainly for small portable lighting outfits, hand torches and telephones. The advent of the radio in the 1920's greatly stimulated the production of dry batteries and the introduction of the transistor radio in more recent times has furthered the growth of carbon-zinc battery production.

As the current drain increases, the carbon-zinc battery is no longer an economic proposition, and although a range of higher powered dry batteries was developed for appliances such as shavers, television sets and tape recorders, it is being displaced by batteries which provide more energy per unit cost.

These batteries are secondary (rechargeable) types in contrast to the carbon-zinc which is a primary type and cannot therefore be charged, so that whilst the initial outlay is higher, battery life can be measured in hundreds of cycles of discharge and charge, compared to the single discharge of the dry battery.

SEALED NICKEL-CADMIUM BATTERIES

The introduction some years ago of the sealed nickel-cadmium battery made available a power source for many cordless appliances for which the carbon-zinc battery was uneconomic.

This type of sealed nickel-cadmium battery uses sintered plates of very thin section so that an element of positive and negative plates interleaved with separators can be packed into a small volume, some of the smallest cells being no larger than a small button.

The plates are made highly porous by sintering carbonyl nickel powder which is then impregnated with the active materials.

The technique of providing a fully sealed cell which can be given many cycles of discharge and charge is to provide negative and positive active materials such that the negative (cadmium) plates have an excess of capacity over the positives (nickel hydroxide).

During charge, gassing occurs first at the positive plate, and the cell is designed so that the oxygen evolved at the positive plate migrates to the negative plate where it is reduced electrochemically. The electrolytic gases which are normally evolved as the plates become charged combine as they are formed to produce water. Providing the charging equipment is correctly matched to the battery size, continuous charging at a low current can take place without the risk of a dangerous rise in internal pressure.

The success of the rechargeable sealed nickel-cadmium battery for cordless appliances has ensured a steady market, particularly in the United States, where in 1969 about 20 per cent of some 7 million electric shavers used nickel-cadmium batteries.

Another popular application for the sealed battery is the electric toothbrush, whilst other growing outlets include electronic photoflash equipment, rural telephone systems, and small emergency lighting systems.

LEAD-ACID BATTERIES

In the past decade new types of maintenance-free and unspillable lead-acid cells have been developed which have a considerable cost advantage over the nickel-cadmium battery. They therefore fill the gap between the carbon-zinc primary battery and the expensive nickel-cadmium battery in providing a reliable source of power for cordless equipment where initial cost is important.

The development of maintenance-free lead-acid batteries was not possible on the same lines as the nickel-cadmium battery and the lead-acid battery is therefore not completely sealed. The electro-chemical reduction of oxygen gas cannot be used since there has to be some slight excess of acid at the plates, which are of limited surface area compared to the plates of the sealed nickel-cadmium battery.

The main problems which had to be overcome were the loss of water from the sulphuric acid during charging, which would eventually cause drying out and battery failure, and spillage of acid and the risk of electrolytic gas escaping from the battery on charge.

All of these problems have been solved by the use of special plate grid alloys of low antimony content or a lead-calcium mixture together with absorbent separators so that little or no free acid is present. The cells are usually vented in such a way that no acid mist escapes, and gassing is kept to a minimum by employing a charger controlled for voltage and current, so that when the battery reaches a voltage approaching gassing point (around 2·35 V per cell), the voltage is maintained at this value and the current tapers off to a safe low value.

In the U.K. and European countries as a whole, cordless appliances such as electric toothbrushes and carving knives have not caught on to the same extent as in the U.S., and the major outlets for sealed rechargeable batteries are emergency lighting units, lanterns and communication systems such as police walkie-talkie sets. On the other hand, cordless lawn mowers powered by lead-acid batteries are quite widely used in the U.K. and total about 250,000. The battery supplied for this equipment is the conventional lead-acid Automotive type, using plates with grids of low antimony content in order to keep open-circuit losses to a minimum.

The mowing season in the U.K. extends over a period of 7 months with the machine in use on average twice a week. The battery is capable of providing power for continuous cutting for about 1 hr, equal to some 800 m² (1000 sq. yd) of turf. Many battery electric mowers have a built-in charger and the battery is recharged on the machine by simply plugging in to a mains supply. The charger is designed to give a low rate output so that when the battery is discharged, the time to fully recharge is about 48 hr.

During the 5-month off-season, it is recommended that the battery be given a refresher charge for about 12 to 24 hr once every 7 or 8 weeks. If electrolyte levels are low following the charge, water should be added and a mixing charge of 3 to 4 hr given.

Battery lives of 5 years or more are common for this application.

WATER-ACTIVATED BATTERIES

World-wide interest is being shown in the development of water-activated lead-acid batteries for use on cars, lawnmowers, etc.

These are dry charged batteries containing the sulphuric acid as a solid gel between the plates of the cells, or in a compressed block of absorbent silica situated above the cell tops. In both cases the batteries are activated by merely adding water.

The elimination of sulphuric acid packed in bottles for batteries sent by mail order or in stock at garages and retailers would be most welcome.

Batteries for Electric Cars

In the U.K. road vehicles and industrial trucks powered by lead-acid batteries are commonplace and reached the total of 50,000 vehicles and 75,000 trucks in 1969. Smaller battery electric vehicles are used throughout the world in airports, golf courses, hospitals and in some industries in the U.S.A. for personnel transportation.

The latest phase is the development of numerous experimental battery electric cars in various countries including the U.S.A., Japan, Sweden, Germany, Italy and the U.K.

The high traffic density, with the attendant congestion and air pollution, in all the large cities, particularly in the U.S.A. and Japan, has stimulated interest at the highest level in the search for alternatives to the internal combustion engine. In the U.S.A. a special committee has been set up by the Senate and money invested in a programme of research on the problems and hazards posed by the generation of something like 11 million cubic metres (400 million cubic feet) of poisonous gas per day from the exhaust pipes of car engines.

The non-pollution characteristics of the battery electric car made it an obvious first choice as a possible solution to the problem, and a vast amount of literature from the countries named above has been issued on the subject. Most of the early prototype cars used lead-acid batteries.

In the U.S., small three-wheel vehicles manufactured by Cushman Motors Inc. were made for the Post Office Department in 1959, and in 1962 E.S.B. Inc. provided larger vehicles.

National Union Electric Corporation converted about 60 Renault Dauphines to battery electric vehicles (named the "Henney Kilowatt").

Electric Fuel Propulsion Inc. converted Renault-10 cars using 96 V 205 Ah lead-acid batteries and 50 cars are on the road for evaluation by various companies. The car has a range of 110–190 km (70–120 miles) for city driving, and a maximum speed of 95 km/hr (60 m.p.h.).

Other American companies who are conducting trials with electric cars powered by lead-acid batteries include E.S.B. Inc., Ford Motor Co., Gould-National Batteries Inc., West Penn Power Co., Westinghouse and General Electric. This last company are using lead-acid batteries as the main energy source with nickel-cadmium batteries operating in parallel to help out during periods of increased load when accelerating or climbing hills.

Some American companies are trying higher power density batteries such as, for example, lithium-nickel fluoride batteries, silver-zinc batteries, sodium/sulphur batteries and hydrogen-oxygen fuel batteries.

In the U.K., various small cars powered by lead-acid batteries are on trial. These include two converted Mini-travellers (Electricity Council), the "Scamp" (Scottish Aviation), the Ford Comuta (Ford Motor Co.), and the Carter Coaster (Carter Engineering). The Enfield 465 (Figs. 9.9 and 9.10), made by Enfield Automotive, Wimbledon, London, is a purpose-built electric car intended for quantity production.

Fig. 9.9. Enfield 465 Battery Electric Car Developed by Enfield Automotive, London

The specifications of the Enfield 465 include:

Motor	48 V d.c. series type, rated at 4·6 h.p. at 2,200 rev/min
Car dimensions	Overall length, 274 cm (108 in.) Overall height, 132 cm (52 in.) Overall width, 127 cm (50 in.)
Performance	Max. speed 65 km/hr (40 m.p.h.) Average range per battery charge 60 km (30–40 miles)
Batteries	Lead-acid 48-V 158 Ah at the 10 hr rate Weight 200 kg (440 lb)
Charger	Integral with car 25 A d.c. output

This car was one of many demonstrated at the first international Electric Vehicle Symposium held in Phoenix, Arizona in November 1969, where in addition to firms from all parts of the U.S. 14 other countries were represented. Of interest was a statement by one of

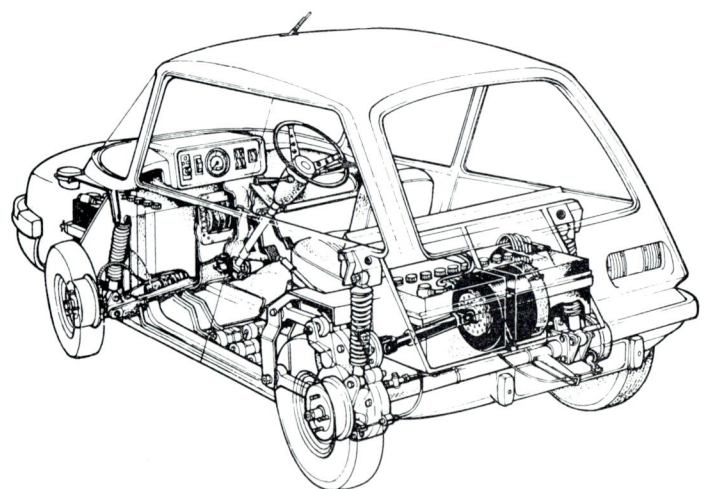

FIG. 9.10. CUTAWAY DRAWING OF ENFIELD 465 SHOWING ELECTRIC MOTOR, TWO BATTERIES, TRANSMISSION AND INTERIOR LAYOUT

the delegates from Japan that 38 prototype electric cars were on trial in Japan.

The maximum speed of 65 km/hr (40 m.p.h.) and 60 km (35 mile) range of the Enfield electric car are typical of cars powered by conventional lead-acid or nickel-cadmium batteries, and to be comparable with the normal passenger car the electric car should have a range of 350 km (200 miles) per battery charge. The other disadvantage of the present battery for the electric car is the time required to fully charge, which is in the region of 5 to 8 hr.

Even these limitations do not rule this type of car for urban use, as over 70 per cent of car owners in the U.K. do not average more than 48 km (30 miles) per day.

It is unlikely that conventional batteries with an energy/density output of 33 Wh/kg (15 Wh/lb) can be improved to provide a range of 350 km or so per charge.

This kind of range might be possible from new electrochemical energy-conversion systems which are at present under investigation and which are described in Chapter 11.

CHAPTER 10

TESTING, FAULT DIAGNOSIS AND REPAIRS

TESTS of the electrical characteristics of batteries are very important to the battery manufacturer in proving production samples and developing new designs.

Other routine tests to check the physical and chemical properties of all the materials used in the various components are also important in maintaining the highest standard of chemical purity that is so essential for the electrochemical efficiency of the battery in service. Tests taken to investigate the electrical characteristics include

(*a*) Capacity tests at all rates of discharge from the 1-min to the 20-hr, at various temperatures.

(*b*) Life tests.

The capacity tests enable the voltage characteristics to be determined and the ampere-hour and watt-hour capacities to be evaluated.

Life tests include conditions that are far more severe than would be experienced in service.

Tests in Service

Tests of batteries in service usually have the following objects—

(*a*) To establish the state of discharge or charge at any time.

(*b*) To ascertain whether the capacity is adequate for the duty, or to detect possible deterioration in performance as the battery ages.

These tests, if taken at regular intervals, are useful in revealing battery troubles (so called) which develop for a variety of reasons not associated with the battery itself. They therefore serve as a useful guide to the operation of the electrical system as a whole.

The state of discharge of a lead-acid battery can easily be determined by measuring the specific gravity of the electrolyte by means of a hydrometer. Readings will, however, be affected by incorrect levels, or the recent addition of water.

HYDROMETER TEST

Most battery manufacturers market a reliable syringe-type hydro-meter consisting of a glass barrel closed at the ends by a rubber bulb and end tube, with a hydrometer float contained in the glass barrel.

To measure the specific gravity of the electrolyte, remove the cell vent plug, squeeze the rubber bulb, and insert the end tube into the

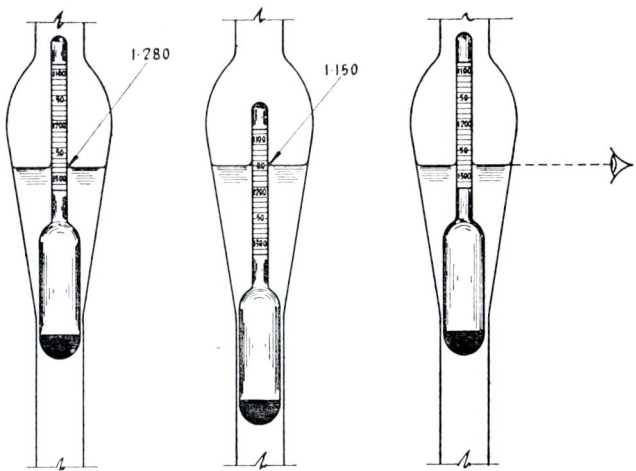

FIG. 10.1.   READING THE HYDROMETER
(a) High float means high specific gravity
(b) Low float means low specific gravity
(c) Correct method of reading hydrometer: eye on level with surface of electrolyte

electrolyte of the cell.   Release the bulb gently and draw just sufficient acid into the glass barrel to float the hydrometer.   Holding the barrel vertical so that the hydrometer floats freely, read the float scale at the acid level.   The acid should be returned to the cell from which it was withdrawn.

The correct method of reading the hydrometer is shown in Fig. 10.1.

The temperature of the electrolyte should also be noted whenever a specific gravity reading is taken.   When this temperature is abnormally high or low, it will be necessary to apply a correction, as the hydrometer reading varies with the temperature of the electrolyte, and must be corrected to a standard temperature of 60° F (15° C).

For all practical purposes this can be done as follows—

For each 10° F *above* 60° F, *add* 0·004 to observed reading
For each 10° F *below* 60° F, *subtract* 0·004 from observed reading

For each 10° C *above* 15° C, *add* 0·007 to observed reading
For each 10° C *below* 15° C, *subtract* 0·007 from observed reading

Examples

1·250 at 70° F corrected becomes 1·254 at 60° F
1·250 at 21° C corrected becomes 1·254 at 15° C
1·250 at 0° C corrected becomes 1·240 at 15° C

RELATIONSHIP BETWEEN SPECIFIC GRAVITY AND STATE OF DISCHARGE

It has been stated in Chapter 3 that there is a linear relationship between the specific gravity of the electrolyte and the ampere-hour output of a battery. The chart shown in Fig. 10.2 relates the approximate state of discharge of most types of battery to the specific gravity reading. Because the working range of specific gravity, for example, 1·280–1·120 for many portable cells, and 1·210–1·150 for many stationary cells, varies with different designs and makes, the chart is intended as only a rough guide.

STATE OF BATTERY FROM OPEN-CIRCUIT VOLTAGE

The linear relationship between the open-circuit voltage of a cell and the specific gravity of the electrolyte (Fig. 10.2) also provides an indication of the state of discharge. This relationship relies on the battery having stood on open-circuit for at least 12 hr following a charge, or about 1 hr following a discharge. The meter used for measuring the cell voltage should be accurately calibrated to read one-hundredths of a volt between about 1·90 and 2·20 V.

BATTERY ON CHARGE

Spot checks of voltage or specific gravity taken infrequently during the recharge of a discharged battery are not very useful guides to battery condition. Charge readings, to be useful, should be taken at regular intervals throughout the complete recharge. Constancy of battery voltage and specific gravity readings, together with gassing of all cells, indicates that the battery is fully charged. The voltage of the battery when fully charged will depend on its age and temperature and the charging current.

For example, an old battery charging at the correct "bench" or finishing rate may be fully charged with a voltage on charge of 2·4–2·5 V per cell. A new battery under similar conditions may be only 80–90 per cent charged when showing 2·4 V, and its voltage, fully charged, may be as high as 2·7 V.

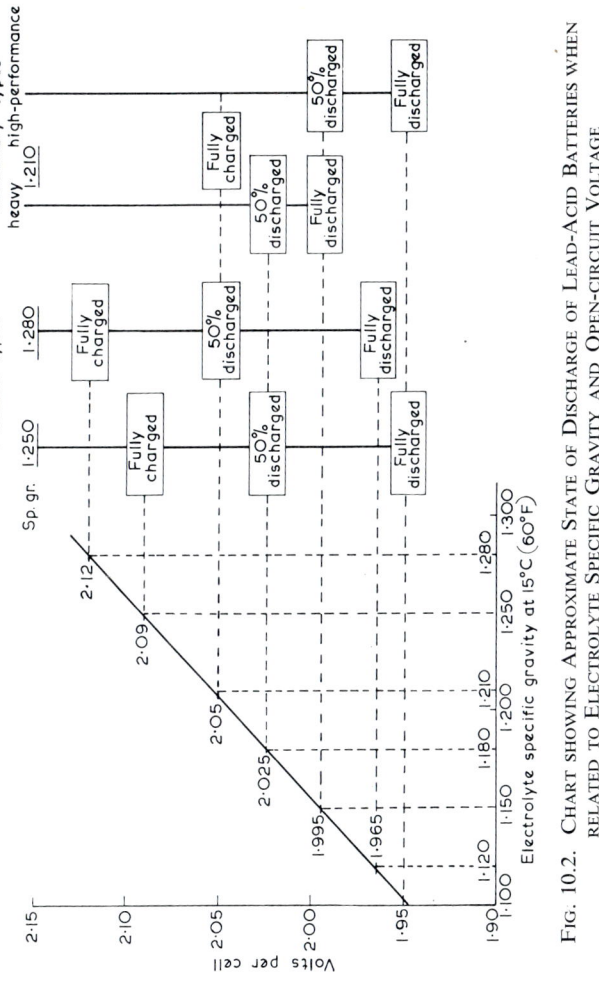

Fig. 10.2. Chart showing Approximate State of Discharge of Lead-Acid Batteries when related to Electrolyte Specific Gravity and Open-circuit Voltage

Where batteries are system controlled and maintained in a charged condition by working in parallel with a charging supply, battery voltage readings may be used as a guide to correction operation. A general indication of the charge characteristics of fully charged cells in good condition is shown in Fig. 10.3.

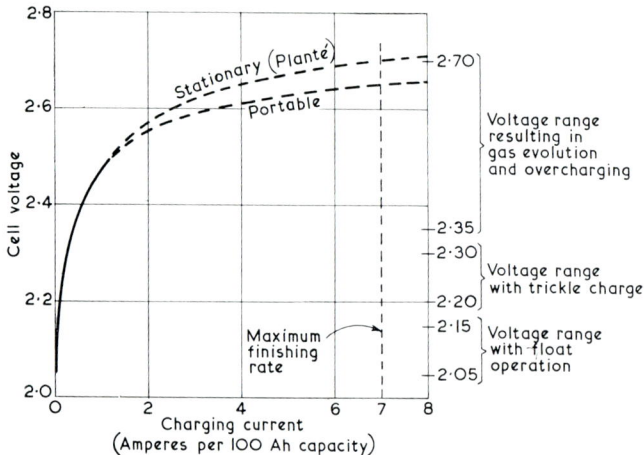

FIG. 10.3. CHARGE VOLTAGE/CURRENT CHARACTERISTICS OF FULLY CHARGED LEAD-ACID CELLS

FLOATING BATTERY

The voltage of a floating battery is usually between 2·05 and 2·15 V per cell. If the voltage is lower than 2·05 V the battery is discharging and will have to be recharged fairly frequently. If it is 2·15 V the battery will require refresher charges about every 3 months to compensate for partial internal losses.

TRICKLE-CHARGED BATTERY

For correct operation the voltage range of a trickle-charged battery should be within 2·2 and 2·3 V per cell. Above 2·3 V there is a danger of the battery being overcharged.

Discharge Tests

In certain installations, as for example in aircraft, it is necessary to test the batteries periodically to ensure reliability at all times.

To carry out a test the battery is fully charged at the specified rate, and allowed to stand for about 6 hr, or until the battery temperature has fallen to between 15° and 25° C, before being placed on discharge. This avoids having to apply a correction for temperature to the capacity obtained on test.

The discharge is usually taken at the 10-hr rate specified by the maker to a final voltage of 1·8 V per cell.

To check a 24-V battery of capacity 25 Ah at the 10-hr rate, the procedure is as follows—

1. Check the electrolyte levels and adjust if necessary by adding pure water to approximately 5 mm above the tops of the separators. Do not overtop.

2. Charge the battery at the recommended current, usually equal to 0·1 × C amperes (where C is the nominal ampere-hour capacity).

3. During the charge, take readings of battery voltage, and of electrolyte specific gravity and temperature at hourly intervals until three consecutive readings are constant and all cells are gassing freely. The charge may take any time between 8 and 24 hr, depending on the condition of the battery.

4. Stop the charge and allow the battery to stand for a period of 6 hr, or longer, if the temperature exceeds 25° C.

5. With the battery temperature within the limits of 15° and 25° C, start the discharge at 2·5 A. Take readings of voltage and temperature 5 min after commencement of discharge and thereafter at 30-min intervals.

6. The discharge is completed when the battery voltage falls to the specified final voltage equivalent to 1·8 V per cell for aircraft batteries (1·75 V for most other portable cells).

If the time of discharge is 8 hr or more, the battery is considered fit for further service. If the time of discharge is less than 8 hr, repeat the test, and if the time is still less than 8 hr, reject the battery.

The meters used for battery testing should be not less than first-grade quality, the voltmeter having two or more ranges so as to be capable of reading overall battery voltage as well as cell voltage. The discharge resistor may be of the slide-wire type rated to carry the current and wattage dissipated by the battery on discharge.

For the case under consideration, the rating of the resistor for 2·5 A at 24 V would be 10 Ω 60 W. It should be capable of adjustment down to about 8 Ω to maintain the current at a steady value as the battery voltage falls during discharge.

HEAVY-CURRENT DISCHARGE TESTS

Where batteries have to be tested at various currents up to several hundred amperes, a circuit similar to that shown in Fig. 10.4 may be used.

The discharge rig can be constructed using different lengths or gauges of ordinary iron wire as resistors. Where 12-V batteries are to be tested, the wires are selected so as to pass the currents indicated at about 10–12 V. The wires should be immersed in water to prevent overheating.

Fine adjustment of the current, in order to maintain a steady value during the test, is obtained by means of a variable carbon-stack resistor in parallel with the wire resistors.

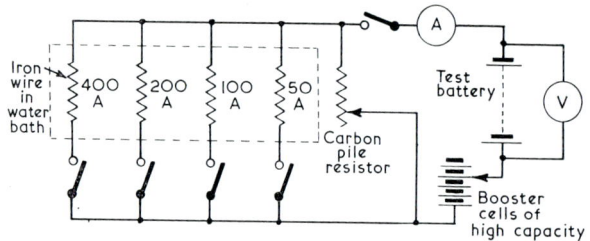

FIG. 10.4. TEST RIG FOR THE CONTROLLED DISCHARGING OF BATTERIES

The booster cells connected in series with the test battery are of large capacity and help to maintain the voltage across the resistors, so that the current can be controlled even for wide variations in the voltage of the test battery during discharge.

By means of this circuit, any test current from about 50 A to 800 A could be obtained by suitable combination of the various resistors.

For discharging batteries of higher voltage than 24 V at currents up to 100 A, when 2 kW or more may be dissipated, a simple water-bath resistor may be used.

The water resistor consists of two metal plates immersed in a tank of water, with the plates connected through a switch across the battery to be discharged. The plates are usually unpasted lead-alloy grids or sheets of lead or iron. A small quantity of sulphuric acid or salt is added to the water to increase its conductivity. The current is controlled by varying the space between the plates, or by adding more acid and adjusting the water level. The plates should not be allowed to touch, as contact between them will short-circuit the battery.

Where considerable power is dissipated in the water-bath resistor, the water will boil, and it is advisable to maintain the level by adding water during the test.

Although somewhat crude, the water-bath resistor with reasonable attention will enable the discharge current to be closely controlled during the course of a test.

Fault Detection and Diagnosis

As with all manufactured articles, battery troubles revealed in service may have one of several origins, but the main sources of trouble are due to some form of abuse. Abuse in service includes inadequate maintenance, incorrect operation, and particularly incorrect charging.

Many so-called battery troubles can be traced to the charging source, and undercharging or excessive overcharging eventually produces battery trouble. The more cells there are in a battery, the greater is the likelihood of slight differences developing in the electrochemical characteristics of individual cells if the battery is somewhat undercharged. To equalize or maintain all cells of the battery in balance requires some slight measure of overcharge during the life of the battery.

The table shown on page 196 is intended to serve as a guide in detecting and correcting trouble in service.

Effect of Impurities in Batteries

Great care is taken to ensure that the materials and the sulphuric acid used in the manufacture of lead-acid batteries are of the highest standard of purity. Impurities which may be present in batteries in service are almost certainly introduced through the use of impure topping-up water, or by corrosion products entering the battery from a corroded terminal or steelwork.

Traction and automotive batteries can tolerate a higher level of contamination than Planté or low-loss cells, which rely on chemical purity for maintaining the low open-circuit losses which are essential for satisfactory operation and life.

METALLIC IMPURITIES

Iron is one of the most common impurities. It is usually introduced by unsuitable topping-up water, but may be present owing to corrosion products getting into the battery. Iron produces discharge of both the positive and negative plates.

It can be detected by chemical analysis of the electrolyte, and by low readings of voltage and specific gravity when the battery has been given a prolonged equalizing charge. Battery condition may be improved by changing the electrolyte several times.

Copper can be present in a battery where copper-reinforced lead posts are used, owing to the presence of a blow hole in the lead post.

Trouble	Symptoms	Probable cause	Remedy
AUTOMOTIVE BATTERIES Sluggish, poor starting performance	Sp. gr. readings uniformly low, or several readings low	Insufficient charge	Give extended equalizing charge. If at end of this, cell voltages on charge are uniformly high and sp. gr. rises to within 10 points of specified value, battery is fit for further service
TRACTION BATTERIES Failure to complete shift or service round	Sp. gr. readings still low after extended equalizing charge	Sulphated plates	Give further charge, and if there is little response and cell voltages on charge are low, check battery capacity
STATIONARY BATTERIES Plates unhealthy colour: positives, light brown; negatives, dark grey		Loss of acid through spillage	Give further charge, and if cell voltages are uniformly high, with no change in sp. gr., add acid to adjust to correct density (seek maker's advice first)
		Broken box, if odd cell low in sp. gr. or showing low level	Replace boxes where level is consistently lower than in remainder
	Several cells showing low charge voltage at end of extended charge	If these are together in one section of the battery, may be tapped section for auxiliary loads	Give these cells a further charge
		Internal short-circuits	Open cells and examine for damage or displaced separators, lead particles between plates, or buckled plates
	Readings of voltage and sp. gr. very erratic after equalizing charge of at least 48 hr	Battery has probably reached end of life	Check 1. Age of battery 2. Capacity 3. Appearance of plates 4. Depth of sediment below plates If capacity is below 80 per cent, plates are in poor condition, or mud space is full, replace battery
Battery overheats	Some connexions very hot	Poor contacts or badly welded joints	Clean and tighten all bolted connexions. Reweld doubtful welded joints
	Battery temperature frequently above 100° F (40° C)	Car battery may be mounted near engine	If possible, direct a stream of outside air round or over battery
	Battery consumes excessive amount of water	Poor ventilation of traction battery during charge. Has possibly been discharged at high rates and/ or overcharged	Check working routine and charging system. Ensure that battery covers are lifted during charge. If possible, blow cool air over cell tops. Check input on charge
Battery damp and dirty, wood trays rotted, or metalwork corroded	Obvious on visual inspection	Bad maintenance, overtopping, or lid sealing compound cracked	Keep battery dry and clean. Do not overtop when adding water. Clean away all traces of acid and old sealing compound from cell lids, and reseal

Copper is deposited on the negative plates, and if present in the battery in sufficient quantity, can cause increased open-circuit losses, lowering of the charging voltage, and loss of capacity.

Analysis of the electrolyte does not provide conclusive evidence of the presence of copper.

Antimony is usually extracted from the grids of automotive and traction batteries by some form of abuse in service. Like copper, it is deposited on the negative plates and causes high open-circuit losses and low top-of-charge voltages. Planté and low-loss batteries do not suffer to the same extent from antimony poisoning, as their grids contain little or no antimony.

Non-metallic Impurities

Chlorine, like iron, is a very common impurity. It is usually introduced in the topping-up water, and can be present in the drinking water supplies of certain areas in amounts which are dangerous for batteries. Chlorides are oxidized, at the positive plate during charge, to perchloric acid, which in time may attack the grids and eventually cause disintegration of the grid structure.

Analysis of the electrolyte may be unhelpful in detecting the presence of chlorides, as the perchloric acid tends to be confined in the positive plates.

Impurities such as petrol, lubricating oil and paraffin are sometimes added in mistake for water. Their presence can usually be detected by the characteristic odour. The impurities tend to coat and blank off the surface of the plates, thus reducing their capacity. Softening of the sealing compound and sometimes of the container may also be evident.

Certain other organic impurities, such as vinegar, methylated spirit and anti-freeze solutions, produce attack and deterioration of the positive plate grids.

Sulphation

Normal sulphation takes place each time a battery is discharged, by the electrochemical reactions between the sulphuric acid electrolyte and the active materials of the plates.

Abnormal sulphation is set up by some form of abuse. The battery may have been undercharged over a period of time, or it may have stood in a discharged condition. The sulphate in the plates forms large crystals which cannot be converted to the active materials during charge.

Frequent deep discharges, below the specific gravity recommended by the battery manufacturer, may produce sulphation trouble due to the difficulty of reconversion to active material during charge. Operating batteries at high temperatures, or with the electrolyte at above normal specific gravity, may produce sulphation troubles. External evidence is given by reduced specific gravity of the electrolyte, and charging the battery does not increase the specific gravity to the correct value owing to much of the sulphate being unconverted.

Internal evidence is apparent in the condition of the plates, particularly the negatives. Sulphated negatives are gritty and harsh in texture, or may have the consistency of toothpaste instead of metallic spongy lead. When the surface of the sulphated negative plate is stroked with the thumb nail, it does not have the metallic sheen of a healthy negative plate.

Many sulphated batteries may be partially restored by emptying out the electrolyte and refilling with pure water, followed by a long charge at the equalizing rate until the specific gravity ceases to rise. Another method is to give the sulphated battery a charge at normal rate for about 48 hr with the electrolyte temperature in the region of 49° C (120° F).

With the above methods the specific gravity after treatment may rise above the specified value, showing that sulphate has been recovered from the plates. The specific gravity should be adjusted to the correct value by adding water.

Repairs

When carrying out repairs to lead-acid batteries it is essential to have suitable tools and equipment to complete the work satisfactorily. The following list of equipment, (*see* Figs. 10.5 and 10.6) is recommended for battery repair work—

1. 12 mm drill, preferably electric (400 rev/min), mounted on a stand so that the drill can be held vertical to the work
2. Centre bit or hollow post cutter
3. Connector puller (for traction batteries)
4. 2 pairs of pliers, for pulling elements out of box
5. Steamer, or heating cap
6. Sealing compound pan
7. Blow-lamp or similar source of heat for warming sealing compound pan
8. Sealing nut spanner (for lock-nut type cells)
9. Putty knife for removing compound
10. Various burning rings (these are placed over parts to be welded to contain the molten lead)

11. Lid puller
12. Antimonial-lead burning stick
13. Burning equipment

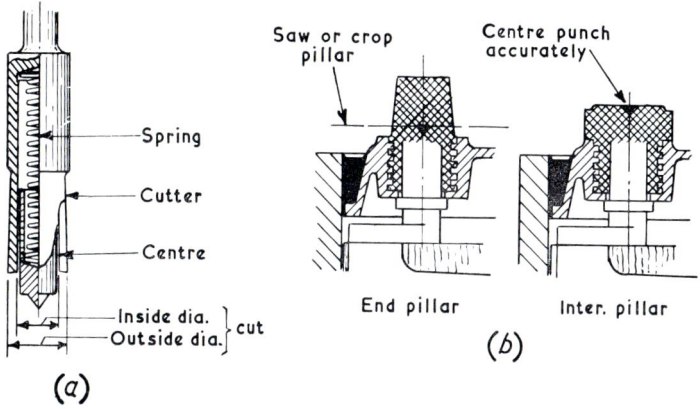

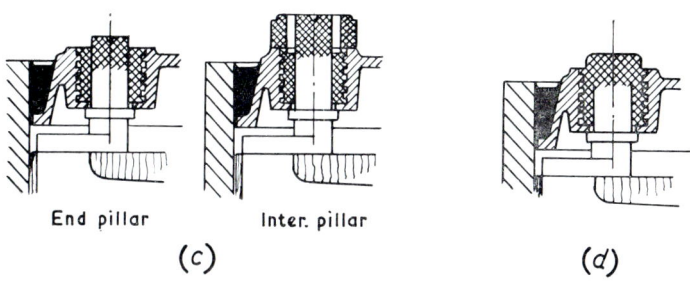

FIG. 10.5. USING HOLLOW POST CUTTER

(a) Typical hollow post cutter
(b) Before using cutter
(c) After using cutter
(d) Lid reburned: showing primary burn made before building up terminals and adding intercell connectors

Lead burning is the term applied to the fusing or welding together of the lead parts by means of a hot flame or carbon rod. In lead burning one of the following methods is used—

(a) Coal gas and oxygen flame
(b) Hydrogen and oxygen flame
(c) Hydrogen and compressed-air flame
(d) Carbon-arc burning equipment

When a good supply of public mains gas is available the first method is used, and to prevent any possibility of a blow-back, a non-return valve is fitted in the supply pipe. The oxy-coal-gas mixture is controlled by a thumb valve in the burner head, and a little experience enables the operator to adjust for the desired flame.

Before attempting any lead burning it is essential that all the lead parts to be welded should be wiped clean and dry, and scraped or wire-brushed to make the surfaces bright.

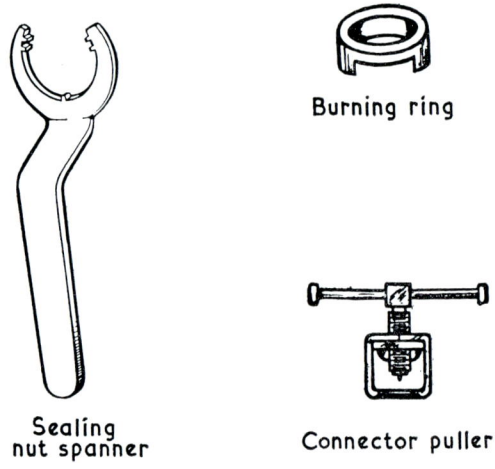

Burning ring

Sealing
nut spanner

Connector puller

FIG. 10.6. ITEMS OF REPAIR EQUIPMENT

To prevent scorching of the lids or container, and particles of lead entering the cells, it is a wise plan to place damp cloths across the battery top. *Air should be blown in the top of the cells before placing the damp cloth in position, as this disperses any gas and avoids the possibility of an explosion.*

CARBON-ARC BURNING EQUIPMENT

The term *arc burning* is a misnomer, as the weld is made by puddling the lead with the white-hot carbon rod, and not by an arc or flame.

The equipment shown in Fig. 10.7 provides a simple and convenient form of lead burning for small jobs, such as the burning of intercell connectors to cell posts. Normally the battery under repair provides the current for heating the carbon rod, or alternatively a separate battery is used. The carbon rod should be moved around over the lead surfaces to keep the lead molten.

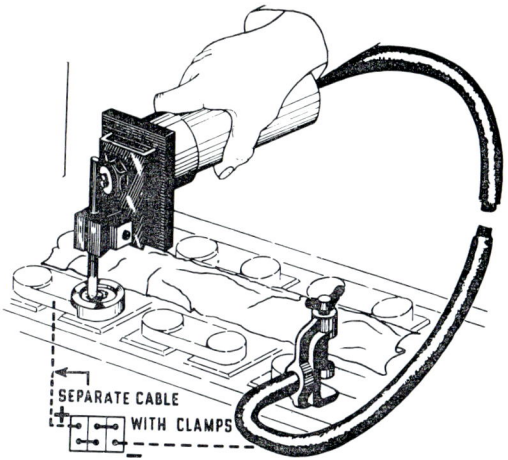

FIG. 10.7. CARBON-ARC BURNING EQUIPMENT
Dotted lines show connexions when separate battery is used

Dark glasses or goggles should be worn as protection against the intense light and the risk of splashing metal.

Repair of Automotive Batteries

COMPLETE REPLATAL

The following procedure should be followed—

1. Note battery layout and dump the acid.
2. Centre punch accurately the intercell connector burns, and gently drill to a depth of 6 mm ($\frac{1}{4}$ in.) only.
3. Prise off the drilled connectors by means of a chisel and retain if in good order; otherwise renew.
4. Soften the sealing compound by placing the battery under a heater cap, or infra-red lamps, or in a steamer. Do not overheat, or the lids will become soft and be distorted when lifted off the battery by the lid puller.
5. Whilst the lids are warm, remove the compound from the edges, using the putty knife, so that they can be used again.
6. Remove the elements from the container compartments. Wash the container carefully, and examine for leaks or damage.

REASSEMBLY

1. Insert new elements and sufficient side packings to ensure a push fit in the container compartments.

2. Check the polarity of all elements for correct assembly in the container.

3. Fit the lids, and ensure they are down by tapping gently.

4. Make a primary burn between each cell post and lid, using the burning ring (Fig. 10.5 (*d*)). The burning ring prevents the molten lead from flowing away from the joint.

5. Seal the lids by pouring molten compound in the lid troughs. Complete the operation by flaming the compound in the lid troughs with a soft flame.

6. Fit the intercell connectors, and using the burning rings, burn the cell post to the connector.

Repair of Traction Batteries

Repairs to traction batteries are usually of a minor nature such as the replacement of a damaged lid or cell box, and more rarely, the replacement of an element.

DISMANTLING THE BATTERY

Before disturbing any part of the battery, make a sketch of its layout and note the polarity of each terminal, as this information will be required when reassembling the cells.

Removing a Cell from the Tray

Use the connector puller by fitting it over the connector and screwing down the spindle until the connector is forced off.

Alternatively, drill the connector to a depth of 6 mm and prise off the connector.

Remove any packings, then lift the cell out of the tray by gripping the cell posts with two pairs of pliers. If the box sticks to the tray or adjacent cells, run a thin-bladed knife between them.

Removing an Element from a Box

Unseal the lid by one of the following methods—

(i) Soften the top of the box with a flame. Run a warm sealing knife through the compound close to the inside of the box walls.

(ii) Place the cell in a steamer or under a heater cap.

To remove the element complete with lid, hold the box firmly between the feet and pull on the cell posts using two pairs of pliers.

If it is necessary to remove the lid from cells fitted with sealing-nut-type lids, unscrew the nuts using the sealing-nut spanner before removing the lid.

For rubber-gasket-seal-type lids, lift the element until there is a 6 mm space between the lid skirt and the top edge of the cell box. Insert two strips of wood, each about 25 mm wide and 5 mm thick, between the lid and cell box, so that the lid is well supported at each end. Apply a downward pressure on the pillars so that the lid comes away with the rubber gaskets.

For a lid with lead inserts, the connector burn is drilled between the lead insert and post, using the hollow post cutter.

Elements, when removed from the boxes, must not be exposed to the atmosphere for long, as the negative plates will oxidize and get very hot. If there is going to be any delay in replacing elements in acid, it is advisable to submerge them in a tank of weak acid until required.

REASSEMBLING THE BATTERY

(*a*) When cell boxes and lids are to be reused, clean off the old sealing compound.

(*b*) Slide the elements in the cell boxes, ensuring a push fit by fitting packing sheets between the end plates and the inside walls of the box.

(*c*) Fit the lids on the cells. For sealing-nut-type lids (Fig. 10.8 (*a*)) fit a rubber washer on each terminal pillar and place the lid in position. Put on the sealing nuts and tighten with the sealing-nut spanner. Lock the sealing nuts by punching the pillar threads at three points around the circumference.

For gasket-seal-type lids (Fig. 10.8 (*b*)) fit the lid on to the cell. Grease the inside bore of a rubber gasket, and press it hard into the recessed cup of the lid. Fill the space above the gasket with warm petroleum jelly, and press the lead post ring into it until the flange locates firmly on the cell lid.

For insert-type lids, (Fig. 10.8 (*c*)), fit the lid on to cell and burn the terminal post to the lid insert, using a primary burning ring.

(*d*) Seal the cell lids using new sealing compound. Pour the compound, heated to about 110° C (230° F), into the trough around the lid, and smooth it over with a mild flame.

(*e*) Assemble the cells in the tray to the original layout, wedging them tightly with thin wood or p.v.c. packings.

(*f*) Reburn the connectors using a burning ring to prevent lead running away from the joint.

Damp cloths or cotton waste should be placed over the vent plugs to avoid the possibility of explosion of the gaseous mixture contained in the upper part of the cell.

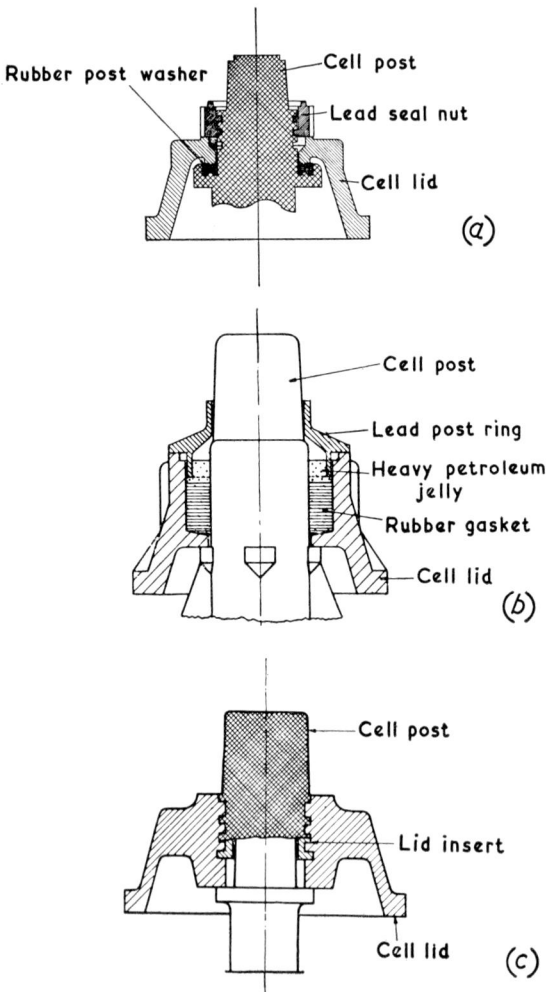

FIG. 10.8. DETAILS OF LID-TO-PILLAR SEALING
(*a*) Sealing nut
(*b*) Gasket seal
(*c*) Insert lid

PREPARING THE REPAIRED BATTERY FOR SERVICE

(*a*) If new uncharged elements have been fitted, the cells should be filled with acid and given a first charge, in accordance with the manufacturer's instructions for new cells.

(*b*) If the original elements were used, the cells should be filled with acid of the same specific gravity as that of the old acid recorded before dismantling, and the cells given an extended charge. At the completion of this charge the specific gravity should be adjusted, if necessary, to the specified value.

Safety

In all battery repair work where acid is handled, suitable precautions should be taken. Protective clothing should be worn, and it is also advisable to protect the eyes by wearing goggles.

Keep all naked flames away from a battery during and shortly after a charge. This is to avoid the risk of an explosion caused by ignition of the gaseous mixture in the tops of the cells and immediate vicinity.

CHAPTER 11

ELECTROCHEMICAL ENERGY DEVICES

Fuel Cells

STORAGE, or secondary, batteries produce electric current by oxidation-reduction, "redox," chemical reactions.

These same chemical reactions take place in fuel cells, but the basic difference between fuel and storage batteries is that in the latter chemical energy is stored in the positive and negative electrodes, whereas in fuel cells the oxidant and the fuel are stored outside the cells, and must be fed to the electrodes continuously during the time the fuel cell is required to supply electric current.

In this respect fuel cells have an advantage over storage cells, since these fuels which are stored outside the cells can quickly be replenished, in somewhat similar fashion to filling up the petrol tank of a car. Storage batteries when fully discharged take several hours to be recharged. However, the high cost of fuel and fuel cell components make them uneconomical for most commercial applications in spite of the optimistic future forecast for them some years ago.

PRINCIPLES

The principle of a hydrogen/oxygen fuel cell was first described by Sir William Grove in 1839. In 1959 F. T. Bacon, after many years of development work, succeeded in designing a fuel battery which delivered sufficient power to drive an industrial forklift truck.

The principle of operation of the hydrogen/oxygen fuel cell is the reverse of the process of electrolysis of water. The two electrodes of the fuel cell are fed with a continuous stream of hydrogen and oxygen (or air), and at the surface of the electrodes the oxygen and hydrogen ions react with the potassium hydroxide electrolyte to produce water.

The overall cell reaction can be simplified as:

$$2 H_2 + O_2 = 2 H_2O$$

The basic reactions taking place in the cell are more fully demonstrated in Fig. 11.1.

The fuel battery has been an essential part of the manned Apollo space missions which successfully put a man on the moon.

For the Apollo programme, on-board power was supplied by the Bacon-type hydrogen/oxygen fuel battery, manufactured by Pratt and Whitney Aircraft Company, whilst other types of hydrogen/oxygen fuel cells made by Allis-Chalmers have provided power in unmanned satellites and space probes.

The development of fuel batteries for this space programme, where performance and reliability combined with low weight were of prime importance, had been in progress for many years and cost

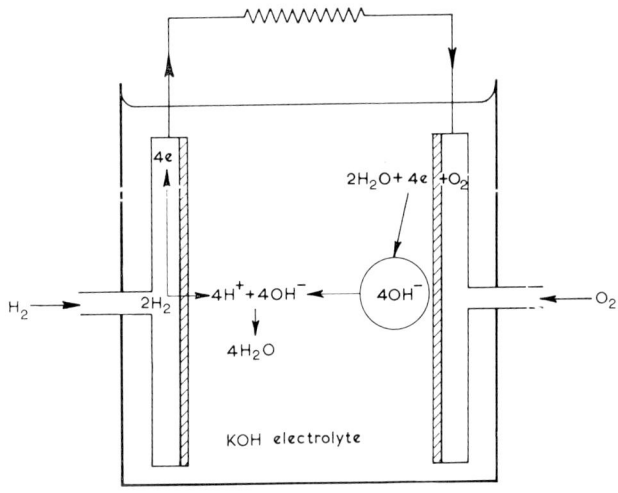

FIG. 11.1. SCHEMATIC PRESENTATION OF HYDROGEN/OXYGEN
FUEL CELL

millions of dollars per annum, to bring the project to a satisfactory conclusion.

The basic principles of the fuel batteries for the Apollo space mission are similar to those used by F. T. Bacon in his hydrogen/oxygen battery.

The electrodes are made from sintered nickel plaques having a coarse pore surface and a fine pore surface, the two surfaces being for gas and electrolyte respectively. The electrolyte is caustic potash of about 85 per cent concentration, whilst the water vapour formed as a by-product of the reaction is removed by condensation from the stream of hydrogen passing over the back of the fuel electrode.

The cells operate at a temperature range of 200–250° C at atmospheric pressure.

Two methods of assembly have been adopted:

1. *Sandwich Assembly*

In this method, two disc type electrodes about 1·8 mm thick, are placed face to face with a thin layer of electrolyte between, and sealed round the periphery by insulating gaskets.

The overall length of a complete cell is about 18 mm and numbers of these cells are stacked together to give any desired working voltage.

At 200–250° C, the voltage on load at a current density of 150 mA/cm^{-2} is about 0·9 V.

For the Apollo mission, the fuel cells were enclosed in a metal container with nitrogen filling the free space in order to preserve the internal atmospheric pressure during flight.

2. *Folded Can Assembly*

Here, a central oxygen electrode is made in the shape of a double-walled can and fitted between two hydrogen electrodes.

As in the sandwich assembly, any system voltage is possible by stacking the required number of cells in series, but in the case of the folded can assembly, cells can readily be removed and replaced.

DEVELOPMENTS IN THE UNITED STATES OF AMERICA

Many companies in the United States have done a considerable amount of development work on fuel cells in the past 10 years. However, in the last 12 months or so, major companies, as for example, Monsanto, Union Carbide and General Electric, have severely pruned their domestic programmes of research.

In addition to Pratt & Whitney Company, one other company in the U.S. which has brought their development work to practical conclusions is Allis Chalmers Manufacturing Company. Using fuel batteries of the type which use hydrogen or hydrazine as fuel and which use sintered nickel electrodes, they have produced batteries to power tractors, fork-lift trucks, golf carts, and an experimental underwater vehicle, Star 1, built by General Dynamics. In addition they have N.A.S.A. contracts to develop auxiliary power systems for the space vehicles.

The General Electric Company has built fuel batteries for the "Gemini" space project, and done a great deal of work of commercial importance in finding ways of using cheap organic fuels, such as natural gas and alcohols. The burning question appears to be whether the promising results obtained on laboratory single electrode cells can be translated into commercial products at an economic cost.

The "TARGET" Programme

This is a private fuel-cell programme financed by about 23 gas companies in the U.S., in conjunction with the Institute of Gas Technology of Chicago. The TARGET (Team to Advance Research for Gas Energy Transmission) system is to reform natural gas with steam, and react the hydrogen in an acid electrolyte fuel cell.

The intention is to use the power from the natural-gas-fed fuel cells to provide a.c. power (d.c. output converted to a.c. by inverters) for a house or a block of buildings. Although the early optimism of providing complete fuel batteries for less than $100 per kW has not been achieved, there is reason to believe that recent technological achievements might produce a fuel battery at about $300 per kW. This would still be very costly compared with power from conventional systems such as engine generators and power stations. However, the encouraging progress made with the TARGET programme would appear to ensure continued development work on fuel cells for this important commercial outlet.

DEVELOPMENTS IN THE UNITED KINGDOM

Three main companies, Shell Research Ltd., Energy Conversion Ltd. and Electric Power Storage Ltd. have been actively engaged on fuel-cell programmes.

1. Shell Research Ltd.

At the Shell Research laboratories, for several years work has been mainly with hydrogen or hydrazine/oxygen fuel cells.

Batteries capable of 5 kW output have been constructed using a novel type of electrode about 0·7 mm thick, made from "Porvic," a microporous P.V.C. material.

The material is coated with a thin layer of silver followed by a noble metal catalyst. These electrodes possess fine pores of uniform distribution, and with the electrolyte space down to 4 mm width it is possible to build a battery of 4 cells per 25 mm of length.

Multi-cell units using porous plastic electrodes, an alkaline electrolyte, and either hydrogen or hydrazine fuel have given outputs of 750 W, with the electrodes operating satisfactorily at 100–150 mA/cm^{-2} at 25–60° C with a cell voltage of 0·7–0·8 V.

2. Energy Conversion Ltd.

This company was formed as a joint consortium, representing the National Research and Development Corporation (N.R.D.C.) and three companies, British Ropes Ltd., British Petroleum Ltd. and

Guest, Keen and Nettlefold Ltd., with the sole object of developing a commercially viable fuel battery.

E.C.L., through N.R.D.C.'s previous association with the work done by Bacon in developing his hydrox battery, have full access to the results and relevant patents of this early research work. They also have licence arrangements with Leesona-Moos Laboratories, and as Pratt and Whitney have used the Bacon-type fuel cell for their space programme, E.C.L. are kept informed on the developments of the Apollo fuel battery system.

3. *Electric Power Storage Ltd.*

The Research and Development Division of this company has carried out work on the development of fuel cells for several years.

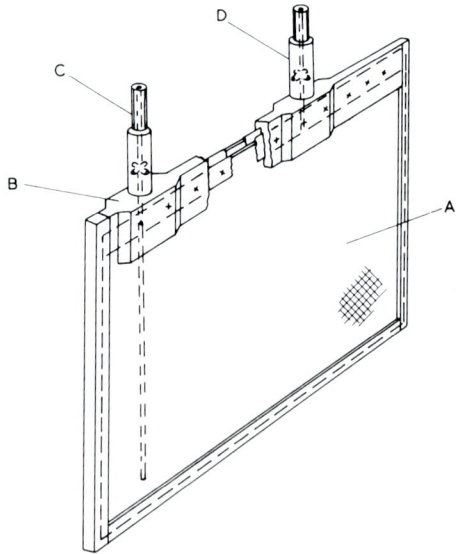

Fig. 11.2. Typical Box Type Electrode

This has included work on various kinds of electrodes, and gaseous and liquid fuels, with the systems operating at low temperature and atmospheric pressure.

The basic principle of plate construction is the same for both positive and negative electrodes. Fig. 11.2 shows a typical electrode consisting of a highly active porous carbon plaque produced by pressing a mixture of powders of carbon, graphite, organic binder and catalyst into a metal gauze. This gauze acts as a conductor for the current and provides mechanical strength and support.

Two such plaques (A) are sealed round the edges to form a hollow electrode (B) and metal terminal posts (C and D) are welded to the top of the plates. The posts, which serve as current take-offs, are hollow to allow the reacting gas to be fed into the electrode gas space.

CELL CONSTRUCTION

The construction of the fuel cell will depend on the application for which it is to be used.

In the hollow electrode cell, a number of positive and negative electrodes are interleaved alternately and the hydrogen and oxygen gas fed to the hollow terminal posts of the negative and positive electrodes respectively. A single cell measuring 152 mm × 172 mm × 89 mm ($6'' \times 6\frac{3}{4}'' \times 3\frac{1}{2}''$) and containing 4 hydrogen and 5 oxygen hollow-type electrodes will deliver 100 A at 0·6 V continuously.

PERFORMANCE

The power density of fuel cells varies between 9 and 31 watts per kg (4 and 14 watts per lb). The voltage and power output of the cell

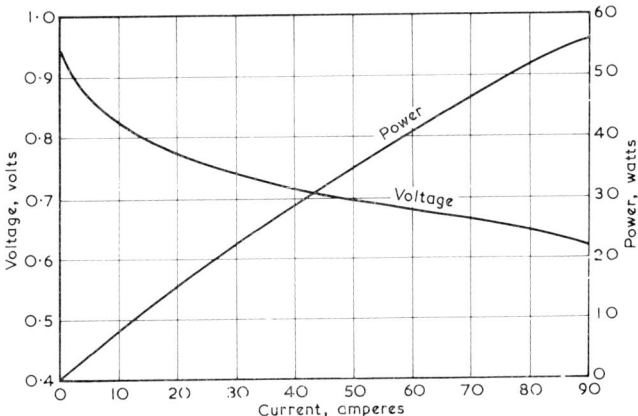

FIG. 11.3. PERFORMANCE CURVES OF A HYDROGEN FUEL CELL

detailed above at various current loads is shown in Fig. 11.3. The energy density of hydrogen/oxygen fuel batteries compared with lead-acid batteries is shown in Fig. 11.4.

The life of the above type of cell has not been fully determined as they have operated in the laboratory continuously for over 20,000 hr without deterioration. Also, a 1 kW battery of these cells has been continuously operated for almost 2 years.

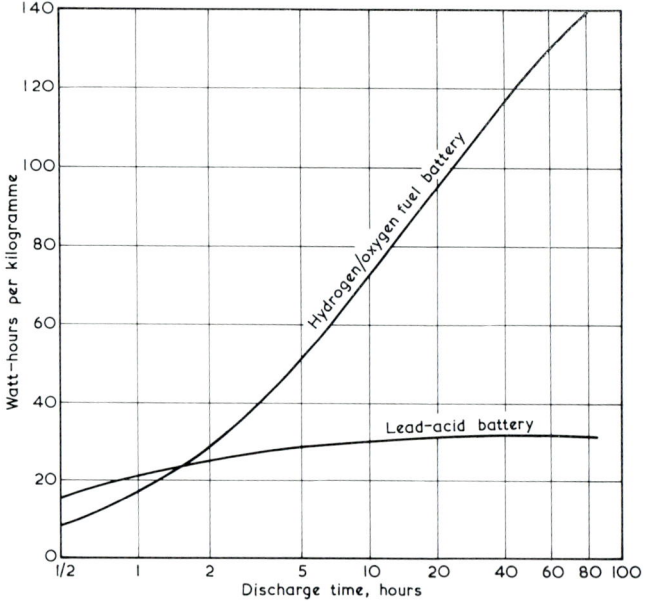

FIG. 11.4. ENERGY DENSITY OF FUEL AND LEAD-ACID BATTERIES

STACK-TYPE FUEL BATTERY

A stack-type fuel battery as shown in Fig. 11.5 offers the advantages of compactness and weight compared to a fuel battery made from the type of cell described previously.

In this design, a number of electrodes are assembled in a stack-type construction, with the reacting gases and the electrolyte fed through channels in the frames of each electrode.

This form of construction eliminates the need for a multiplicity of pipe connections between each cell and larger batteries can be easily constructed.

In both designs, the electrolyte is a 30 per cent solution of potassium hydroxide, which as the cell operates, is gradually diluted by the water produced from the cell reaction.

The cells are designed to enable the electrolyte to be circulated to an outside reservoir or tank where the concentration can be

FIG. 11.5. STACK-BUILT FUEL BATTERY

adjusted, either by evaporation or by addition of concentrated potassium hydroxide. A typical schematic arrangement of a fuel battery system is shown in Fig. 11.6.

The fuel battery consists of a series of electrodes which must be supplied with fuel and oxidant. The hydrogen and oxygen (when used) are usually stored in cylinders under compression. In the diagram shown, air is used as the oxidant under forced circulation, and the carbon dioxide is and must be removed from the air by passage through a "scrubber" before passing into the cells.

The water produced during the reaction is removed by circulation into the KOH tank or by evaporation outside the cell. The operating

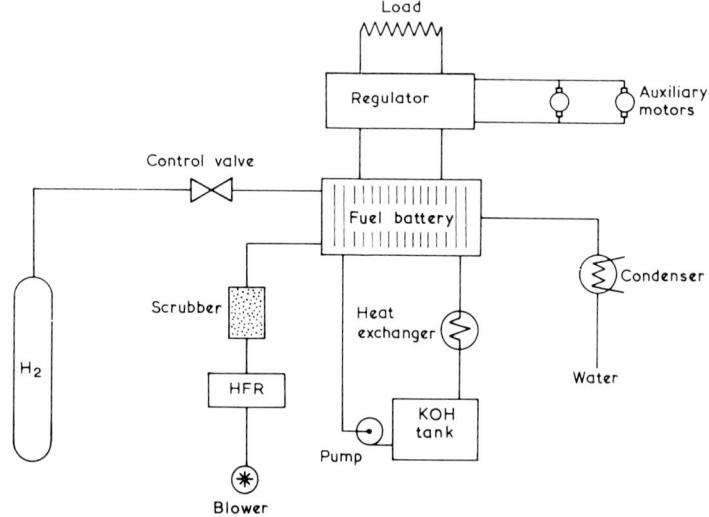

FIG. 11.6. SCHEMATIC DIAGRAM OF FUEL BATTERY SYSTEM

temperature of the battery is controlled by means of the heat exchanger in the circulation system.

APPLICATIONS

Fuel-cell systems can be used to advantage for applications where electrical energy is required for long periods. Such applications include:

Trucks for handling industrial materials
Road and rail traction
Naval craft and submarines
Navigational aids (buoys and beacons)
Radio repeater stations

At Electric Power Storage Ltd. (E.P.S.) an industrial truck has been powered by fuel cells and operated for several years in the laboratories. The truck with fuel battery is shown in Fig. 11.7. The battery consists of 63 hydrogen/oxygen fuel cells operating as a 48-V system, and supplying a maximum power of 3·5 kW.

The supply of hydrogen and oxygen gas is contained in gas cylinders located beneath the fuel battery. There is sufficient gas for an 8-hr working period, and replacement cylinders can be fitted

FIG. 11.7. FUEL BATTERY POWERED TRUCK

in less than 5 min. This enables the truck to work a multiple-shift system.

Fig. 11.8 shows a 60-W navigational beacon powered by a special type of hydrogen/oxygen fuel battery to provide exceptional reliability. The beacon runs for 14 hr every day of the year. The 600 kWh fuel battery requires attention every 6 months only, when some electrolyte must be drained from the cells and the concentration adjusted. This type of equipment has been in continuous operation in the E.P.S. laboratories for almost 2 years.

FUTURE PROSPECTS

It is obvious from what has been achieved so far that commercial fuel batteries can be made for various applications. However, the big obstacle in their development is cost. This applies to the materials used, the types of catalyst and the nature of the fuels.

Fig. 11.8. 600-kWh Fuel Battery Operating Navigational Beacon

As regards fuels, hydrogen is by far the most widely used. If used directly as compressed gas it involves the use of heavy cumbrous metal cylinders, whilst indirect production of hydrogen involves cracking/reforming units to break down hydrocarbon fuels.

There are also engineering problems involving pumps, heat exchangers, condensers and reforming units.

The market prospects for fuel batteries appear to be brightest in the field of low-maintenance applications such as navigational buoys, beacons, and repeater stations, where reliability and freedom from attention over-ride cost considerations.

Metal/Air Batteries

The need for energy density outputs greater than those available from conventional storage batteries of the lead-acid or nickel-cadmium types for such applications as electric cars has opened up a field of research into more reactive anode materials. As in the fuel-cell development, the other line of attack has been to use oxygen (air) electrodes for the cathode, thus using air which is virtually free and imposes little or no weight penalty.

There are two main types of metal/air batteries:

(a) Secondary or rechargeable types, in which the anode material is reconstituted by passing in a (charging) current.

(b) Mechanically rechargeable batteries, where discharged anodes are removed and replaced with new ones.

ZINC-AIR CELLS

This is the most common type of metal/air cell. In the secondary rechargeable form the zinc hydrate produced at the anode during discharge is reconverted to zinc metal on recharge, whilst in the mechanically rechargeable form new zinc electrodes are fitted at the end of each discharge.

Recharging zinc-air cells has, in the past, presented considerable difficulties due to short circuits developing in the cell caused by dendrite growth of the zinc. Work on this problem has been done in various countries, and it is reported that the problem has been partially solved by techniques developed by the Edison Electric Institute and Gulf Atomic Inc. They have designed a 20 kW (laboratory) battery, comprising 40 cells assembled as eight 5-cell modules. Each module includes ancillary equipment for gas circulation, and electrolyte pumps.

The positive electrodes are porous nickel sandwiches which are supplied with oxygen. The zinc negative electrodes are removed from the modules for charging; the operation of replating the zinc taking about three hours. It is estimated that the battery will have an energy capacity of 25 kWh for a weight of 270 kg (600 lb). This would be over twice the capacity of a lead-acid battery of similar weight, and if used for electric car propulsion should provide a range between charges of about 75 miles.

Here again there are doubts about such batteries being commercially viable due to uncertain life and high cost.

The mechanically rechargeable zinc-air battery, where discharged anodes are replaced with new ones, has proved of considerable value in the Services although expensive and inconvenient for many commercial applications.

ALKALI-METAL BATTERIES

Batteries with the highest theoretical energy and power densities would use anodes of alkali metals such as sodium, lithium and potassium, which are the most reactive of metals.

High-temperature alkali metal batteries are really the only source of energy and power for electric car propulsion which would most nearly match that of the I/C engine as regards range, power and acceleration.

One of the more promising developments in this field is the use of sodium/sulphur reactants at operating temperature of 300° C with a potential energy density output of more than 400 Wh/kg (200 Wh/lb).

Development programmes using these reactants are in hand at Ford Motor Company (U.S.A.) and British Rail Technical Centre, Derby (U.K.). It is hoped that the system will eventually develop into a cheap source of power for electric trains and electric vehicles. In the meantime very formidable problems related to choice of materials, engineering design, and safety will have to be solved before this type of battery can be a commercial proposition or accepted as a safe-risk by the prospective customer.

GENERAL COMPARISON WITH LEAD-ACID BATTERIES

Concentrated efforts involving the vast expenditure of money and man-power over the past 10 years or so have been successful in producing electrochemical systems which have higher energy/density outputs than conventional lead-acid or nickel-cadmium batteries.

In the case of fuel-cell systems of the hydrogen/oxygen type, commercial batteries have been produced but they are usually confined to special applications as, for example, the Apollo spacecraft. It is likely that because of the high cost of component parts and the relatively bulky and heavy fuel supply systems their use will continue to be limited to their present outlets. Similarly metal/air batteries with energy/density outputs much greater than those of lead-acid batteries are satisfying some commercial, but mainly Service, applications.

It has to be remembered that the improved energy output of the above systems is only possible at relatively low discharge current densities and both fuel and metal/air batteries provide relatively low power density. This is where the high-temperature alkali-metal systems would score if the development work now proceeding in various countries could produce a commercially acceptable battery. Before this can happen a very large number of problems must be solved, and in the meantime conventional storage batteries such as the lead-acid type will continue as a most reliable source of stored electrical energy.

Developments in lead-acid battery manufacture will continue to produce some improvement in output per unit weight by reductions in top lead (smaller intercell connectors), use of lighter container materials, and more efficient plate active materials. From the operation point of view, servicing will be simplified by the introduction of maintenance free batteries, improved charge acceptance and safer charging (automatic) systems.

APPENDIX 1

WATER AND ACID FOR USE IN LEAD-ACID BATTERIES

LEAD-ACID batteries rely on a high standard of purity of materials used in their manufacture for successful performance and long life. It is very important that the sulphuric acid used for first filling of the battery, and the water used for topping-up purposes during the life of the battery, should conform to the standard of purity recommended by the battery maker.

Water

Only pure water of the standard approved by the battery maker should be used for diluting sulphuric acid or for topping-up purposes.

The regular use of impure water results in a build-up of impurities within the battery, since they remain there whilst the water is removed by electrolysis as hydrogen and oxygen gas.

DOMESTIC WATER

This is water available from the mains supply, and commonly known as tap water.

In many parts of the United Kingdom, domestic water is suitable for topping-up batteries, but there are also many supplies which are impure for batteries, the impurity content varying from season to season. It is important that domestic water should be approved by the battery maker before being used for topping-up.

DISTILLED AND DEIONIZED WATER

Water produced by still or ion-exchange equipment is suitable for topping-up, but the latter should be checked frequently to ensure that the required standard of purity is maintained.

IMPURITY LIMITS

The water should be clear and colourless when viewed through a column 30 cm deep. It should be odourless and neutral to litmus.

The proposed addition of sodium fluoride (1 part per million) to mains water to retard dental decay will have no harmful effects on batteries.

The usual accepted standard of purity for water used for topping-up purposes restricts the impurities to the following limits.

Impurity	Maximum Concentration (parts per million)
Chlorine	15
Copper	10
Iron	10
Ammonia	10
Arsenic	3
Manganese	0·1
Nitrates and nitrites	10
Total fixed residue	250

Sulphuric Acid

The standard of purity of sulphuric acid used for lead-acid batteries is detailed in British Standard No. 3031 : 1958.

This standard sets out very clearly the maximum permissible impurity levels and gives test methods for determining them.

MIXING AND DILUTING SULPHURIC ACID

When it is necessary to mix or dilute sulphuric acid the vessels or tanks used should be of porcelain or glass, or should be lead lined. Plastic and ebonite boxes may be used where the acid concentration does not exceed 1·400 sp. gr.

When mixing *always add acid to water.* The acid should be poured on the water slowly because heat is generated as the liquids meet.

Care should be taken to avoid splashes as acid burns can be dangerous.

The approximate quantities of water required to dilute concentrated acid are given below.

Initial specific gravity	Final specific gravity	Approximate proportions by volume of	
		Acid	Water
1·840	1·400	1	1½
	1·250	1	3½
	1·230	1	4
	1·200	1	4½
	1·170	1	5½
1·400	1·280	2	1
	1·250	3	2
	1·230	1	1
	1·200	3	3½
	1·170	1	1½

APPENDIX 2

QUESTIONS AND ANSWERS

Q. (1) Explain the differences between modified-constant-voltage and constant-current methods of charging storage batteries.

(2) A battery of 35 cells, having a capacity of 100 Ah at the 10-hr rate, in a fully discharged condition, is recharged at a constant current of 10 A from a 110-V d.c. supply. If the cell voltages at the start and end of the charging period are 2·1 and 2·6 V respectively, and the ampere-hour efficiency is 90 per cent, calculate (*a*) time of charge, and (*b*) maximum and minimum values of charging resistance.

(3) The same battery is to be recharged from a modified-constant-voltage supply of 110 V d.c. If the fixed resistance is determined at a cell voltage on charge of 2·5 V so that the current at this voltage is 8 A, calculate (*a*) value of fixed resistance, (*b*) charging current at start and end of charge, and (*c*) time of charge if mean voltage of battery on charge is equivalent to 2·4 V per cell.

A. (1) A full explanation of the two methods is given in Chapter 4.

Briefly, the constant-current method involves almost continuous control of the charging current (by variable resistance, or transductor device, etc.), so that adjustments are made manually or automatically to compensate for rising battery voltage during charge. Without this adjustment, variations in battery or supply voltage would result in corresponding changes in the charging current. In constant-current charging by means of manual control, the variable resistance is set initially at a suitable value to give the desired charging current. The resistance is gradually reduced in order to maintain constant current as the battery voltage increases during charge. The example worked out below illustrates the method. It should be noted that, in all constant-current charging, the current when the battery commences to gas must not exceed 7 to 10 per cent of the ampere-hour capacity (7 to 10 A per 100 Ah).

Modified-constant-voltage charging involves charging from a constant-voltage supply with a fixed value of ballast resistance incorporated in the charging circuit. The value of resistance is chosen so that a safe current flows into the battery towards the completion of the charge, that is, once the battery commences to gas. This current is somewhere between 7 and 10 A per 100 Ah of battery capacity at a mean gassing voltage equivalent to 2·5 V per cell. The charging current falls during charge, and usually the initial charging current is higher, and the final current lower, than that used for the constant-current method of charging.

(2) (*a*) *Time of charge*. The output is 100 Ah, and the efficiency is 90 per cent. The input is therefore 100/0·9, or 111 Ah.

The charging current is constant at 10 A. The charging time is therefore 111/10, or 11·1 hr.

(*b*) *Charging resistance.*

At start of charge,

> Battery voltage = 35 × 2·1 = 73·5 V
> Voltage to be dropped = 110 − 73·5 = 36·5 V
> Maximum resistance = 36·5/10 = 3·65 Ω

At end of charge,

> Battery voltage $= 35 \times 2\cdot6 = 91$ V
> Voltage to be dropped $= 110 - 91 = 19$ V
> Minimum resistance $= 19/10 = 1\cdot9\ \Omega$

(3) (*a*) *Fixed resistance.*

> Battery voltage $= 35 \times 2\cdot5 = 87\cdot5$ V
> Current $= 8$ A
>
> Fixed resistance $= \dfrac{110 - 87\cdot5}{8} = 2\cdot8\ \Omega$ (approx.)

(*b*) *Charging current.*

At start of charge,

> From (2) (*b*), battery voltage $= 73\cdot5$ V
>
> Charging current $= \dfrac{110 - 73\cdot5}{2\cdot8} = 13$ A

At end of charge,

> From (2) (*b*), battery voltage $= 91$ V
>
> Charging current $= \dfrac{110 - 91}{2\cdot8} = 6\cdot8$ A

(*c*) *Charging time.*

> Mean voltage on charge $= 35 \times 2\cdot4$ V
>
> Mean current $= \dfrac{110 - (35 \times 2\cdot4)}{2\cdot8} = 9\cdot3$ A
>
> From (2) (*a*), input $= 111$ Ah
> Charging time $= 111/9\cdot3 = 12$ hr (approx.)

Q. A 12-V lead-acid car battery gives a voltage across its terminals of 9·5 V when supplying 150 A for 10 sec for starting an engine. Immediately the load is removed the battery voltage rises to 12·0 V.

(*a*) Calculate the internal resistance of the battery, and of each cell.

(*b*) What would be the battery voltage when supplying a load of 250 A for several seconds?

A. (*a*) *Internal resistances*

> Resistance of battery $= \dfrac{\text{Voltage drop in battery}}{\text{Current}}$
>
> $= \dfrac{12\cdot0 - 9\cdot5}{150}$
>
> $= 0\cdot0167\ \Omega$

A 12-V lead-acid battery consists of 6 cells in series, so that the resistance per cell is $0\cdot0167/6 = 0\cdot00278\ \Omega$.

(*b*) *Voltage on load.*

> Terminal voltage $=$ Nominal voltage $-$ Voltage drop in battery
> $= 12 - (250 \times 0\cdot0167)$
> $= 7\cdot825$ V

Q. Is there any risk from the gases given off towards the end of charge?

A. With any storage battery the gases given off towards the end of charge are oxygen and hydrogen. The gases are harmless as far as breathing them is concerned, but they can cause an explosion if ignited by a flame or spark. Explosions will not normally occur if the simple precautions of good ventilation, absence of any kind of flame or spark during charge, and for some time after the end of charge, are observed.

Q. Low cell voltages of a discharging lead-acid battery may be due to (*a*) its being quite healthy but in a discharged state, or (*b*) its being reasonably charged but unhealthy, or even at the end of its useful life.

Explain fully how to distinguish between these two possibilities, quoting all suitable values of voltage and specific gravity.

N.B. A high-rate discharge tester should not be used for this purpose, and marks will be deducted if this instrument is mentioned. (C. and G., Motor Vehicle Electricians' Work. 1961.)

A. The student has to quote suitable voltages and specific gravities, so obviously the diagnosis of battery condition entails use of an accurate 0–3 voltmeter, and a syringe-type hydrometer capable of reading 1·100 to 1·300 sp. gr.

(*a*) *Battery healthy but in a discharged state.* In this condition and with the battery on discharge, say at 2 or 3 A (through the side and tail lights), cell voltages would be uniformly low. Typical voltages for a 12-V (6-cell) battery would be as shown in col. (i) of the following table.

Cell No.	(i) Battery partly discharged		(ii) Battery almost completely discharged	
	Volts per cell	Sp. gr.	Volts per cell	Sp. gr.
1	1·8	1·150	1·7	1·120
2	1·75	1·145	1·75	1·120
3	1·8	1·150	1·5	1·110
4	1·7	1·140	1·7	1·115
5	1·8	1·150	1·6	1·110
6	1·75	1·145	1·7	1·120

If, however, the battery were almost completely discharged, it is possible that one or two cell voltages would be somewhat lower than the remainder, as in col. (ii).

In both examples, the readings indicate no significant difference in cell to cell condition but merely that the battery is discharged and requires a charge. Obviously in either of these conditions the battery should be taken off the car or vehicle and given a bench charge, as normal charging on the car will not restore the desired state of charge.

Confirmation that the battery was healthy would be provided by the fact that all cells would readily accept charge. This means that, with the charging current on and the battery in the fully charged condition, all cells would gas freely, and cell voltage and specific gravity readings would be uniformly high.

(*b*) *Battery reasonably charged but unhealthy, or at the end of its useful life.*
Any wide variation in readings, particularly of specific gravity, can usually be
regarded as signifying one of three possibilities: (*a*) a fault in the battery,
(*b*) battery in an unhealthy condition, or (*c*) battery has reached the end of its
useful life.

Typical readings of a 12-V (discharging) battery, which is reasonably charged
but unhealthy, would be as follows—

Cell	1	2	3	4	5	6
Voltage	1·95	2·0	1·85	1·80	1·95	1·80
Sp. gr.	1·250	1·265	1·210	1·200	1·255	1·200

Thus there would be a considerable variation in cell readings, with the 3
readings underlined much lower than the remainder. The readings of the 3 best
cells indicate that the battery is reasonably charged but the condition of the
battery as a whole is poor owing to the three unhealthy cells. A complete bench
charge, if given, might not bring up the weak cells, thus confirming that they
were unhealthy.

Prolonged charging might help to restore the weak cells by raising their voltages
and specific gravities to the level of the good cells. In this event the battery
could be put back into service.

A battery at the end of its useful life would give readings more out of step
than in the previous example—

Cell	1	2	3	4	5	6
Voltage	1·70	1·80	1·60	1·85	1·50	1·80
Sp. gr.	1·180	1·210	1·130	1·220	1·120	1·220

If given a prolonged bench charge, cell voltage and specific gravity readings on
charge would be low, and there would be considerable variation from cell to cell.
This would confirm that the battery had reached the end of its useful life.

Q. What is a counter-e.m.f. cell?

A. This is a "gas" cell, in which the gases formed at the surfaces of the electrodes produce a voltage in opposition to the applied voltage. The lead-acid
counter-e.m.f. cell consists of blank antimonial lead grids in sulphuric acid
electrolyte. The alkaline counter-e.m.f. cell consists of nickel-plated iron plates
in caustic potash electrolyte. These cells have negligible capacity and are designed
to carry the maximum load current, as they are connected in series with the load
but in opposition to the main battery, as in Fig. A.1.

The application shown is typical of control schemes used in small telephone
exchanges where a constant voltage is maintained across the load irrespective of
the changes in battery voltage depending whether it is on charge or discharge.
The voltage drop of about 10 V across the counter-e.m.f. cells would be fairly
constant for wide variations in the load current, and would compensate for the
high battery voltage during charge.

The counter-e.m.f. cells would be switched in circuit with the battery on charge,
and out of circuit as the battery voltage fell on discharge.

Q. Is the electricity from a storage battery similar to that from the public
mains supply?

A. An electric current is a continuous flow of electrons in a conductor or circuit, produced by the influence of an electromotive force. The source of electromotive force may be an alternator, a direct-current generator or a storage battery, and they all produce a flow of electricity when connected to a conductor.

The alternator, however, produces electricity which changes direction many times each second. This type of electricity, available from public mains supply, is called *alternating-current* electricity. The storage battery produces unidirectional, or *direct-current*, electricity.

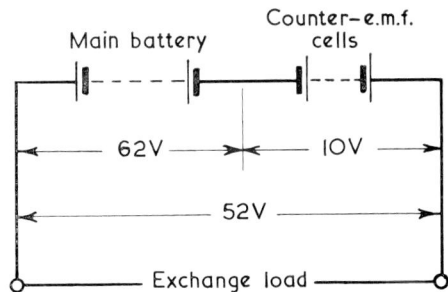

Fig. A.1. Principle of Counter-e.m.f. Cells

Q. Is there danger of electric shock from a storage battery?

A. If the battery voltage is sufficiently high, a person can get an electric shock, particularly if the ends of the battery are touched one with each hand. With wet floors and damp hands an appreciable shock can be received from a battery of 50 V or even lower.

Usually a shock from a.c. mains supply is more dangerous as the person touching the supply is subjected to $\sqrt{2}$ times the nominal voltage. On a 240 V nominal a.c. supply the maximum, or peak, voltage is therefore 339·4 V. At the same time a 240-V battery is capable of rising to 315 V when charging at the normal rate, so that the danger of receiving a shock from a battery is greater when it is on charge than when it is on discharge.

Q. Does topping up with impure water cause explosions?

A. No—but the impurities present in the water will in time affect the working of the battery and may eventually cause failure.

Q. The level of the electrolyte in storage batteries gradually falls during service. Why is this? Why should water only be added to restore the level? What other factors affect the level of electrolyte in a battery?

A. A storage battery, even of the automotive type where there is restricted access to the outside air by way of the vent plugs, will show a fall in electrolyte level both when standing idle and during normal service. Some of this fall in electrolyte level is due to loss of water by evaporation, but with a battery standing idle for several weeks, most of it is due to the release of entrapped gas in the plates and the electrolyte into the atmosphere.

When a battery is cycling in service, that is, alternately discharged and charged, as in traction service, water in the electrolyte is decomposed into hydrogen and oxygen gases towards the end of each charge. This is the gassing action which takes place in a battery as it becomes fully charged. The loss of water causes a progressive fall in electrolyte level and an increase in acid concentration.

When the level falls below the top of the separators or to the recommended minimum, water should be added to restore the electrolyte to the correct level. As the level rises on charge, and particularly at the end of charge due to displacement of the electrolyte by the vigorous gassing taking place, when adding water to a discharged battery it is advisable to allow for this electrolyte "expansion" on charge, and not to add too much water at a time.

Q. Describe the full procedure to be followed in deciding that a lead-acid car battery should be renewed. (C. and G., Motor Vehicle Electricians' Work.)

A. The procedure for diagnosis of the condition of a car battery is described very fully in Chapter 6. The main points are restated below.

Check the battery condition, using simple equipment consisting of a hydrometer, a high-rate discharge tester or a voltage check tester. The main object of any test using these instruments is to see whether there are any cell-to-cell irregularities.

If the readings obtained are regular, even though at a low level, it is almost certain that the battery is reasonably healthy but in a discharged condition.

Irregular readings, where one or more cells are much lower than the remaining cells, nearly always indicate that the battery is reaching the end of life. On the other hand, a battery which has operated on a car in a discharged condition for some time usually has one or more cells somewhat weaker than the remainder. To distinguish the "healthy" discharged battery from the "worn out" discharged battery, it is necessary to give a full bench charge, using a charger which gives an output of about 3 to 5 A, with the battery on charge for about 24 hours, or until the cells are gassing vigorously. Readings of voltage and specific gravity taken at the end of the charge should be reasonably uniform, with voltages per cell in the range between 2·5 and 2·7 V (depending on the age of the battery), and specific gravities between 1·270 and 1·290 (with correct electrolyte levels). If the cells which were noted as being weaker than the remainder, on initial check test, fail to gas or to come up in voltage or specific gravity after a complete charge, it can be assumed that the battery is worn out.

More expensive equipment than that described, such as the Crypton tester model AD43, may also be used. This involves discharging, charging and retesting the battery following a short charge. This type of equipment is very useful and accurate for quick battery diagnosis.

A car battery which is suspect and is known to be 3 years old or more, or has been used for more than 65,000 km, can usually be regarded as approaching the end of its service life.

Q. (*a*) What is the lowest voltage to which it is safe in practice to discharge a lead-acid cell? What is the maximum voltage this cell will reach when fully charged and still on charge?

(*b*) Sketch the curve showing how the voltage of a lead-acid cell varies with time during charge from the lowest safe state of discharge to a fully-charged state.

(C. and G. Motor Vehicle Electricians, Work, 1960.)

A. (*a*) The minimum voltage on discharge is determined more by the nature of the duty than the characteristic of the cell.

For example, when a battery is used for an engine-starting duty where a current of several hundred amperes may be supplied, the minimum voltage required for turning the starter motor is usually taken as 8 V for a 12-V battery, or 1·33 V per cell.

It would do the battery no harm to take it down to a voltage even below 1 V per cell, as at such a high rate of discharge the battery would be by no means exhausted at that point. What does exhaust the battery to the full extent of its capacity is a discharge at a low rate, as, for example, on lighting loads where there is a drain of only 3 or 4 A. The safe voltage limit at light loads is usually taken as 1·75 V per cell, or 10·5 V for a 12-V battery. Prolonging the discharge much beyond this point would produce fairly rapid voltage collapse, as it would be beyond the knee of the discharge voltage characteristic, which in any case is usually regarded as the minimum useful voltage from the cell.

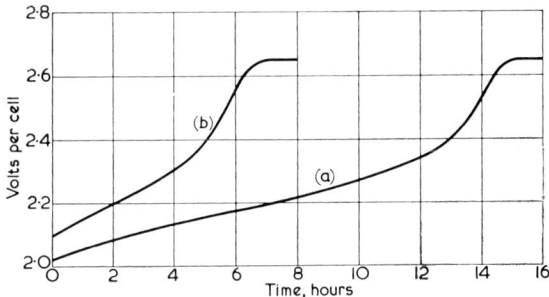

FIG. A.2. TYPICAL RECHARGE-VOLTAGE CHARACTERISTICS OF LEAD-ACID BATTERY AT NORMAL CHARGE RATE

(*a*) Battery 100 per cent discharged (20-hr rate)
(*b*) Battery 50 per cent discharged (20-hr rate)

The maximum voltage attained on charge is usually taken as some value between 2·6 and 2·7 V per cell. It could, however, be outside this range, depending on charging current, age of cell, and temperature of cell.

A charging current above or below that recommended for the cell would produce a corresponding increase or decrease of the charging voltage. An old cell would have a lower charging voltage than a new one. A hot cell would have a lower voltage than a cool cell.

All these variations in charging voltage relate to the cell in the fully charged condition and still on charge.

(*b*) Assuming that a car-type battery is being considered, the lowest safe state of discharge, from the battery point of view, would be that obtained by a low-rate discharge to the specified end voltage, namely the 20-hr rate to an end voltage equivalent to 1·75 V per cell.

From the operational point of view, a car battery in service should not normally be discharged beyond the state where it is no longer reliable for engine starting. This state of discharge would be the equivalent of 40 to 50 per cent of the 20-hr capacity taken from the battery.

Fig. A.2 shows the voltage/time curves of a battery when recharged at normal rate (7 A per 100 Ah) following discharges to the extent of 100 and 50 per cent of the 20-hr capacity.

INDEX